CALVERT MATH

Calvert Math is based upon a previously published textbook series. Calvert School has customized the textbooks using the mathematical principles developed by the original authors. Calvert School wishes to thank the authors for their cooperation. They are:

Audrey V. Buffington
Mathematics Teacher
Wayland Public Schools
Wayland, Massachusetts

Alice R. Garr
Mathematics Department Chairperson
Herricks Middle School
Albertson, New York

Jay Graening
Professor of Mathematics
  and Secondary Education
University of Arkansas
Fayetteville, Arkansas

Philip P. Halloran
Professor, Mathematical Sciences
Central Connecticut State University
New Britain, Connecticut

Michael Mahaffey
Associate Professor,
  Mathematics Education
University of Georgia
Athens, Georgia

Mary A. O'Neal
Mathematics Laboratory Teacher
Brentwood Unified Science
  Magnet School
Los Angeles, California

John H. Stoeckinger
Mathematics Department Chairperson
Carmel High School
Carmel, Indiana

Glen Vannatta
Former Mathematics Supervisor
Special Mathematics Consultant
Indianapolis Public Schools
Indianapolis, Indiana

ISBN-13: 978-1-888287-74-5

Copyright © 2008, 2009 by Calvert School, Inc.

All rights reserved. No part of this book may be reproduced or transmitted in any form or by any means, electronic or mechanical, including photocopying, recording, or by any information storage and retrieval system, without permission in writing from the publisher.

Printed in the U.S.A.
1 2 3 4 5 6 7 8 9 10   12 11

CHIEF LEARNING OFFICER
Gloria D. Julius, Ed.D.

SENIOR CONSULTANT/PROJECT COORDINATOR
Jessie C. Sweeley, *Associate Director of Research and Design*

CURRICULUM CONSULTANT
Robin A. Fiastro, *Manager of Curriculum*

PROJECT FACILITATOR/LEAD RESEARCHER
Hannah Miriam Delventhal, *Math Curriculum Specialist*

FIELD TEST TEACHERS The following Calvert teachers used the Calvert Math prototype in their classrooms and added materials and teaching techniques based on Calvert philosophy and methodology.

Andrew S. Bowers
James O. Coady Jr.
Roman A. Doss
Shannon C. Frederick
Julia T. Holt

Willie T. Little III
John M. McLaughlin
Brian J. Mascuch
Mary Ellen Nessler
Margaret B. Nicolls

Michael E. Paul
Diane E. Proctor
Patricia G. Scott
E. Michael Shawen
Ada M. Stankard

Virginia P.B. White
Jennifer A. Yapsuga
Andrea L. Zavitz

CALVERT EDUCATION SERVICES (CES) CONSULTANTS The following CES staff worked with the Day School faculty in developing the mathematical principles for Calvert Math.

Eileen M. Byrnes
Nicole M. Henry

Linda D. Hummel
Christina L. Mohs

Kelly W. Painter
Mary-Louise Stenchly

Ruth W. Williams

REVISION AUTHORS The following Day School faculty and CES staff contributed to the correlation of the math materials to national standards and to the authoring of revisions based on the correlations.

Nicole M. Henry
Julia T. Holt
Barbara B. Kirby

Willie T. Little III
Mary Ellen Nessler
Hannah Miriam Delventhal

Mary Ethel Vinje
Jennifer A. Yapsuga

EDITORS
Anika Trahan, *Managing Editor*
Eileen A. Baylus
Holly S. Bohart
Anne E. Boszievich

Bernadette Burger
Danica K. Crittenden
Pamela J. Eisenberg
Sarah E. Hedges

Maria R. Kerner
Mary Pfeiffer
Megan L. Snyder

GRAPHIC DESIGNERS
Lauren Loran, *Manager of Graphic Design*
Caitlin Rose Brown
Steven M. Burke
Vickie M. Johnson
Bonnie R. Kitko

Vanessa Ann Panzarino
Deborah A. Sharpe
Joshua C. Steves
Teresa M. Tirabassi

To the Student

Being successful in math depends on having the right tools, skills, and mental attitude. Your math book is a tool to help you discover and master the skills you will need to become powerful in math. These thinking skills will prepare you to solve everyday problems.

About the Art in this Book

Calvert homeschooling students from all over the world and Calvert Day School students have contributed their original art for this book. We hope you enjoy looking at their drawings as you study mathematics.

CONTENTS

Diagnosing Readiness

Diagnostic Skills Pretest xiv

DIAGNOSTIC SKILLS
1. Greater Numbers xvi
2. Rounding xviii
3. Adding xx
4. Column Addition xxii
5. Subtracting xxiv
6. Multiplying by 1-Digit Numbers xxvi
7. Related Facts and Missing Factors .. xxviii
8. Mental Math Computation: Division Patterns xxx
9. Division with Remainders xxxii
10. Dividing 2-Digit Numbers xxxiv
11. Dividing 3-Digit Numbers xxxvi
12. More Dividing xxxviii
13. Fractions Show Parts of a Whole and Parts of a Group xl

Diagnostic Skills Posttest xlii

1 Numbers and Place Value

1.1 Renaming Whole Numbers 2
    MIND BUILDER: History 3
1.2 Millions and Billions 4
1.3 Comparing and Ordering Whole Numbers 6

CUMULATIVE REVIEW 9

1.4 Rounding Whole Numbers 10
    MIND BUILDER: Logical Reasoning 11
1.5 Problem-Solving Strategy: Using the Four-Step Plan 12

Mid-Chapter Review 13

1.6 Decimal Place Value 14
    MIND BUILDER: Attic Numerals 15
1.7 Comparing and Ordering Decimals 16
1.8 Rounding Decimals 18
1.9 Metric Units of Measurement 20

PROBLEM SOLVING
    Checkmate 23

Chapter 1 Review 24
Chapter 1 Test 26

CHANGE OF PACE
    International Standard Notation of Time 27
Cumulative Test 28

v

2 Addition and Subtraction

- 2.1 Properties of Addition 30
- 2.2 Mental Math: Computing Sums and Differences 34
- 2.3 Mental Math: Estimating Sums and Differences 36
- 2.4 Subtracting Greater Numbers 38

PROBLEM SOLVING
- Happy Holidays..................40

CUMULATIVE REVIEW 41
- 2.5 Problem-Solving Strategy: Guess and Check..................42
- Mid-Chapter Review 43
- 2.6 Adding Decimals 44
- 2.7 Subtracting Decimals 46
- 2.8 Problem-Solving Strategy: More Than One Solution..................48
- 2.9 Elapsed Time..................50
- 2.10 Problem-Solving Strategy: Exact Answer or Estimate..................52
  - **MIND BUILDER:** Patterns 53

- Chapter 2 Review 54
- Chapter 2 Test..................56

CHANGE OF PACE
- Roman Numerals 57
- Cumulative Test..................58

3 Multiplication

- 3.1 Properties of Multiplication 60
- 3.2 Mental Math: Multiplying Multiples of 10, 100, and 1,000 62
- 3.3 Mental Math: Estimation Using Rounding 64
- 3.4 The Distributive Property..................66
  - **MIND BUILDER:** Logical Reasoning 68

PROBLEM SOLVING
- Please Don't Squeeze the Oranges..................69
- 3.5 Multiplying by Multiples of 10 and 100 70
- Mid-Chapter Review 71
- 3.6 Multiplying by 2-Digit Numbers 72
- 3.7 Multiplying by 3-Digit Numbers 74
- 3.8 Problem-Solving Strategy: Find a Pattern 76
  - **MIND BUILDER:** Number Puzzles 77
- 3.9 Multiplying Decimals by 10, 100, and 1,000 78
- 3.10 Multiplying Whole Numbers and Decimals..................80

CUMULATIVE REVIEW 83
- 3.11 Multiplying Decimals 84

- Chapter 3 Review 86
- Chapter 3 Test..................88

CHANGE OF PACE
- Exponents 89
- Cumulative Test..................90

4 Division

4.1	Mental Math: Compatible Numbers	92
	MIND BUILDER: Operation Path	94

PROBLEM SOLVING

	Sum Fun	95
4.2	Zeros in the Quotient	96
4.3	Dividing Greater Numbers	98
4.4	Problem-Solving Application: Interpreting Remainders	100
	MIND BUILDER: Sums	102

CUMULATIVE REVIEW 103

4.5	Mental Math: Dividing by Multiples of 10	104

Mid-Chapter Review 105

4.6	Dividing by Multiples of 10	106
4.7	Two-Digit Divisors	108
4.8	Dividing 4- and 5-Digit Dividends	110
4.9	Dividing Decimals by 10, 100, and 1,000	112
4.10	Converting Metric Units	114
4.11	Dividing Decimals by Whole Numbers	116
4.12	Problem-Solving Strategy: Eliminating Possibilities	118
	MIND BUILDER: Estimation	119

Chapter 4 Review 120
Chapter 4 Test 122

CHANGE OF PACE

Computation: Short Division 123
Cumulative Test 124

5 Fractions

5.1	Divisibility Rules	126
5.2	Prime and Composite Numbers	128
5.3	Prime Factorization	130
5.4	Common Factors and Greatest Common Factor	132
5.5	Equivalent Fractions	134
5.6	More Equivalent Fractions	136
5.7	Simplest Form	138
5.8	Problem-Solving Strategy: Choose the Method of Computation	140

Mid-Chapter Review 141

5.9	Mixed Numbers and Improper Fractions	142

CUMULATIVE REVIEW 145

5.10	Renaming Improper Fractions as Mixed Numbers	146
5.11	Renaming Mixed Numbers as Improper Fractions	148
5.12	Fractions as Decimals	150

Chapter 5 Review 152
Chapter 5 Test 154

CHANGE OF PACE

Tangram Activity 155
Cumulative Test 156

6 Adding and Subtracting Fractions

- 6.1 Adding and Subtracting Like Denominators 158
- 6.2 Answers in Simplest Form 160
- 6.3 Least Common Denominators 162
- 6.4 Compare and Order Fractions 164
  - **MIND BUILDER:** Mental Images 165
- 6.5 Adding Fractions with Unlike Denominators 166
  - **MIND BUILDER:** Logical Reasoning 167
- 6.6 Subtracting Fractions with Unlike Denominators 168
- 6.7 Adding and Subtracting Unlike Denominators 170

PROBLEM SOLVING
  Galactic Twins 172
CUMULATIVE REVIEW 173
- 6.8 Problem-Solving Strategy: Relevant Information 174
- Mid-Chapter Review 175
- 6.9 Adding and Subtracting Mixed Numbers 176
- 6.10 Renaming to Subtract 178
- 6.11 Customary Measurement 180
- 6.12 Computing with Measurement 182

Chapter 6 Review 184
Chapter 6 Test 186
CHANGE OF PACE
  Patterns 187
Cumulative Test 188

7 Multiplying and Dividing Fractions

- 7.1 Multiplying Whole Numbers and Fractions 190
- 7.2 Multiplying Fractions 192
  - **MIND BUILDER:** Logical Reasoning 194
CUMULATIVE REVIEW 195
- 7.3 Multiplying a Mixed Number by a Fraction and by a Whole Number 196
- 7.4 Problem-Solving Strategy: Logical Reasoning 198
- Mid-Chapter Review 199
- 7.5 Multiplying Mixed Numbers 200
- 7.6 Dividing a Whole Number by a Fraction 202
- 7.7 Dividing Fractions 204
- 7.8 Problem-Solving Strategy: Solve a Simpler Problem 206

Chapter 7 Review 208
Chapter 7 Test 210
CHANGE OF PACE
  Optical Illusions 211
Cumulative Test 212

8 Geometry

8.1	Geometry Basics	214
8.2	Angles	216
Cumulative Review		219
8.3	Triangles	220
8.4	Polygons	224
8.5	Quadrilaterals	226
8.6	Problem-Solving Strategy: Make a Drawing	228
Mid-Chapter Review		229
8.7	Circles	230
Problem Solving		
	Pick a Pattern	233
8.8	Congruence	234
8.9	Symmetry	236
8.10	Transformations	238
8.11	Three-Dimensional Figures	240
Problem Solving		
	Puzzler	243
Chapter 8 Review		244
Chapter 8 Test		246
Change of Pace		
	Tessellations	247
Cumulative Test		248

9 Perimeter, Area, and Volume

9.1	Perimeter of Polygons	250
	Mind Builder: Perimeters	251
9.2	Circumference	252
9.3	Area of Rectangles	254
9.4	Perimeter and Area of Complex Figures	256
9.5	Problem-Solving Strategy: Draw it	258
9.6	Area of Parallelograms	260
Problem Solving		
	The Great Divide	263
9.7	Area of Triangles	264
Mid-Chapter Review		266
Cumulative Review		267
9.8	Problem-Solving Strategy: Using Venn Diagrams	268
9.9	Three-Dimensional Figures and Nets	270
9.10	Surface Area	272
9.11	Volume	274
9.12	Problem-Solving Application: Using Perimeter, Area, or Volume	276
Chapter 9 Review		278
Chapter 9 Test		280
Change of Pace		
	Drawing Figures and Perimeter	281
Cumulative Test		282

10 Statistics, Graphing, and Probability

10.1	Data and Statistics	284
10.2	Mean, Median, and Mode	286
10.3	Problem-Solving Application: Using Statistics	288
10.4	Stem-and-Leaf Plots	290
10.5	Bar Graphs	292
10.6	Line Graphs	294
10.7	Histograms	296
10.8	Problem-Solving Application: Choose the Proper Graph	298

Mid-Chapter Review 299

PROBLEM SOLVING

    Cross That Bridge When You Come to It 300

CUMULATIVE REVIEW 301

10.9	Graphing Ordered Pairs	302
	MIND BUILDER: Sets	303
10.10	Probability	304
10.11	Experimental Probability	306
10.12	Counting Outcomes	308
10.13	Using Statistics to Predict	310

Chapter 10 Review 312
Chapter 10 Test 314

CHANGE OF PACE

    Means, Medians, and Magic Squares 315

Cumulative Test 316

11 Ratios and Percents

11.1	Ratios	318

CUMULATIVE REVIEW 321

11.2	Equivalent Ratios	322
	MIND BUILDER: Equivalent Ratios	323
11.3	Rates	324
11.4	Problem-Solving Strategy: Choose a Strategy	326

Mid-Chapter Review 327

11.5	Scale Drawings	328
11.6	Percents	330
11.7	Fractions, Decimals, and Percents	332
11.8	Mental Math: Finding 10% of a Number	334
11.9	Percent of a Number	336
11.10	Circle Graphs	338

PROBLEM SOLVING

    Get a Strike 341

11.11	Problem-Solving Application: Discounts	342

Chapter 11 Review 344
Chapter 11 Test 346

CHANGE OF PACE

    Scientific Notation 347

Cumulative Test 348

12 Looking Ahead

12.1 Integers and the Number Line 350
12.2 Expressions and Variables 352
12.3 Mental Math: Solving Equations 354
12.4 Order of Operations......................... 356
12.5 Problem-Solving Strategy:
       Work Backward 358
Mid-Chapter Review 359

PROBLEM SOLVING
   What a Deal! 360
CUMULATIVE REVIEW 361
12.6 Functions...................................... 362
12.7 Graphing Functions 364
12.8 Problem-Solving Strategy:
       Write an Equation......................... 366
       MIND BUILDER: Gears 367

Chapter 12 Review 368
Chapter 12 Test.................................... 370
CHANGE OF PACE
   Fibonacci Numbers 371
Cumulative Test................................... 372

APPENDIX 373
GLOSSARY 375
INDEX 385

Contents **xi**

Diagnosing Readiness

In this preliminary chapter, you will take a diagnostic skills pretest, practice skills that you need to review, and take a diagnostic skills posttest. This preliminary chapter will ensure that you have the skills necessary to begin Chapter 1.

Diagnostic Skills Pretest

Complete the diagnostic skills pretest below. If you get one or more problems in a section incorrect, please refer to the section number in parentheses for review. After reviewing this chapter, take the diagnostic skills posttest at the end of the chapter.

Write the value of the underlined digit. (Section 1)

1. 3<u>8</u>,942
2. 403,2<u>4</u>6
3. <u>3</u>4,895
4. <u>2</u>04,004

Write in standard form. (Section 1)

5. forty-eight thousand, six hundred five
6. thirty thousand, seven hundred twenty-five
7. 300,000 + 4,000 + 6
8. 80,000 + 200 + 90 + 1

Round to the underlined place-value position. (Section 2)

9. 1,<u>7</u>82
10. <u>3</u>,953
11. 8<u>2</u>6
12. <u>9</u>60

Add. (Section 3)

13. 4,357 + 2,954
14. 2,229 + 4,992
15. $23.05 + $2.48
16. $345.08 + $23.79

Add. (Section 4)

17. 36 + 74 + 25
18. 24 + 37 + 46
19. 352 + 503 + 275
20. 3,296 + 4,125 + 85

Subtract. (Section 5)

21. 409 − 87
22. $4.00 − 0.79
23. 712 − 95
24. 840 − 197

Multiply. (Section 6)

25. 85 × 4
26. 402 × 7
27. 8,493 × 6
28. $10.40 × 5

Write the fact family for each set of numbers. (Section 7)

29. 7, 9, 63
30. 7, 8, 56
31. 36, 4, 9
32. 21, 3, 7

Divide. (Section 8)

33. 6)420 34. 4)2,000 35. 4)320 36. 9)270

Divide. (Section 9)

37. 5)26 38. 8)43 39. 9)70 40. 3)11

Divide. (Section 10)

41. 4)89 42. 4)65 43. 3)95 44. 7)84

Divide. (Section 11)

45. 6)714 46. 3)$5.04 47. 4)647 48. 8)9,713

Divide. (Section 12)

49. 9)357 50. 5)425 51. 7)$5.15 52. 7)506

Write a fraction to name the colored part of each whole or each set. (Section 13)

53.   54.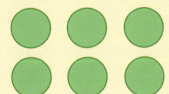

Complete. (Section 13)

55. $\frac{1}{4}$ of 8 = ■   56. $\frac{2}{3}$ of 12 = ■

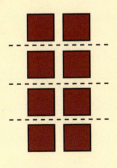

1 Greater Numbers

Alaska has the largest area of any state in the United States. The number in square miles that represents its area is written below in standard form and words.

Standard Form   586,412

Words   five hundred eighty-six thousand, four hundred twelve

Standard form is the way you usually see and work with numbers. The symbol in each place is a digit. The digits are 0, 1, 2, 3, 4, 5, 6, 7, 8, and 9.

The place-value chart shows the value of each digit.

Thousands Period			Ones Period		
hundred thousands	ten thousands	thousands	hundreds	tens	ones
5	8	6 ,	4	1	2

▶ Greater numbers are separated into periods by a comma. Name the periods on the chart.

The value of each digit in 586,412 can be used to express 586,412 in expanded form.

Expanded Form   500,000 + 80,000 + 6,000 + 400 + 10 + 2

More Examples

A. Write in standard form, expanded form, and words.

Standard Form   30,324
Expanded Form   30,000 + 300 + 20 + 4
Words   thirty thousand, three hundred twenty-four

B. Make the number greater by the given amount.

927,085 + 1,000 = 928,085

TRY These

Write the value of the underlined digit.

1. 78<u>5</u>,326
2. 6<u>3</u>9,241
3. 70<u>8</u>,562
4. 400,3<u>2</u>0
5. 75,<u>3</u>00
6. <u>6</u>04,206
7. <u>1</u>7,892
8. 796,43<u>7</u>

Exercises

Write in standard form, expanded form, and words.

1. 321,650
2. forty-one thousand, two hundred nine
3. 600,000 + 2,000 + 400 + 30 + 8
4. 400,325
5. six hundred seventy-eight thousand, five
6. 300,000 + 20,000 + 1,000 + 600 + 50

Make each number greater by the given amount.

7. 45,620 + 1,000
8. 28,937 + 100
9. 7,090 + 10
10. 59,295 + 1,000
11. 41,068 + 100,000
12. 92,370 + 10,000

Answer.

13. Use the digits 5, 1, 8, 4, and 7 to write the greatest and least possible numbers.

1 Greater Numbers

2 Rounding

The men who signed the Declaration of Independence had to travel long distances. The distance from Boston to Philadelphia is 291 miles. To tell someone *about* how far this is, you can round 291 to the nearest ten.

To round a number is to find its approximate value. A rounded number is an estimate of an exact value.

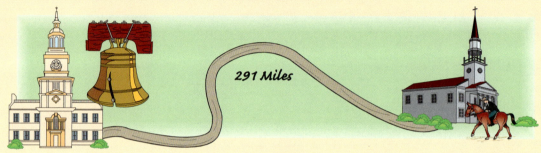

On a number line, you can see that the nearest ten is 290. To the nearest ten, 291 rounds to 290.

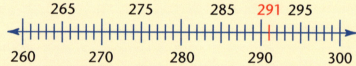

Remember When a number is halfway, round to the greater number.

More Examples

A. To the nearest hundred, 271 rounds to 300.

B. To the nearest thousand, 7,500 rounds to 8,000.

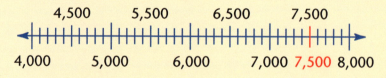

TRY These

Copy the number line and mark the given number. Then round to the nearest ten.

1. 534

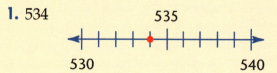

2. 793

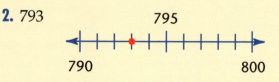

Exercises

Round to the nearest ten.

1. 582 2. 594 3. 565

Round to the nearest hundred.

4. 920 5. 878 6. 1,050 7. 1,028 8. 1,125

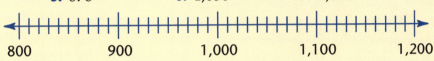

Round to the nearest thousand.

9. 2,500 10. 2,262 11. 5,082 12. 3,627

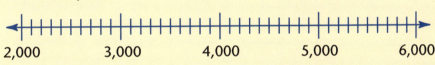

13. Susan's school has 436 students. Round 436 to the nearest ten. Round 436 to the nearest hundred.

14. The distance from Houston to Los Angeles is 1,538 miles. What is the mileage rounded to the nearest thousand?

15. Jerome has $7.39. To the nearest dollar, how much money does he have?

16. The Golden Gate Bridge is 8,981 feet long. Round the length to the nearest thousand.

3 Adding

There are 3,652 books in the Madison School Library. The Stevens School Library has 2,918 books. To find the number of books in the two libraries, add 3,652 and 2,918.

Estimate.
4,000 + 3,000 = 7,000

Step 1	Step 2	Step 3

Add the ones.
Add the tens.
 	
    1
  3,6**52**
+ 2,9**18**
———
     70
 	
Rename 10 ones as 1 ten 0 ones.
	Add the hundreds.
	
	  1 1
	  3,**6**52
	+ 2,**9**18
	———
	   570
	
	Rename 15 hundreds as 1 thousand 5 hundreds.
		Add the thousands.
		
		1 1
		3,652
		+ **2**,918
		———
		6,570

> Be sure to line up the digits by place-value position.

There are 6,570 books in the two libraries. Is the sum close to the estimate?

Check your answer. Reverse the order of the addends.

  2,918
+ 3,652
———
  6,570

More Examples

A.
  1 1
  8,632
+ 3,986
———
12,618

B.
    1
    436
+ 2,900
———
  3,336

C.
   1 1 1
 $98.89
+   1.11
———
$100.00

Add dollars and cents like whole numbers. Place the dollar sign and decimal point in the answer.

TRY These

Add.

1.   6,385
  + 2,176

2.   2,496
  + 5,175

3.   4,873
  + 6,214

4.   6,381
  + 5,217

5.   8,395
  + 2,441

6. $86.39
  + 5.78

Exercises

Add.

1. $62.45 + 83.57

2. 12,765 + 43,594

3. $468.53 + 76.92

4. 93,487 + 65

5. 27,938 + 48,562

6. $43.92 + $2.68

7. 8,891 + 20,790

8. What is the sum of 276,312 and 19,244?

Problem Solving

9. Use the data on the previous page. About how many more books are in the library at Madison than at Stevens?

10. Cheryl earned $28,529 in her business. During the same period, Percy earned $19,730. Together how much did they earn?

Solve. Use the chart.

11. What is the distance from Atlanta to Jacksonville and back?

12. The Lees plan a trip from Los Angeles to Atlanta. Is it shorter to go through Denver or St. Louis?

Air Distances (in kilometers)

	Atlanta	Chicago	Denver	Jacksonville	Los Angeles	St. Louis
Atlanta		963	1,948	445	3,119	781
Chicago	963		1,463	1,387	2,806	413
Denver	1,948	1,463		2,350	1,350	1,257
Jacksonville	445	1,387	2,350		3,450	1,221
Los Angeles	3,119	2,806	1,350	3,450		2,544
St. Louis	781	413	1,257	1,221	2,544	

4 Column Addition

Mrs. Villa asked Alonso to find the total number of books. Alonso counted the books on each shelf. To find the total, he can add.

32 + 46 + 28 + 65 = ?

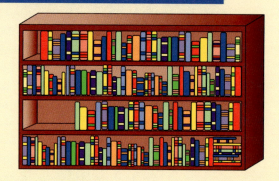

Estimate.
30 + 50 + 30 + 70 = 180

Step 1	Step 2
Add the ones.	Add the tens.
²32 46 28 + 65 ⎯⎯ 1 Record 21 ones as 2 tens 1 one.	²32 46 28 + 65 ⎯⎯ 171 Record 17 tens as 1 hundred 7 tens.

There are 171 books on the bookshelf.

Compared to the estimate, the answer seems reasonable.

More Examples

A. ¹34
   78
+ 21
⎯⎯
133

B. ¹¹257
   408
+ 362
⎯⎯⎯
1,027

C. ¹¹¹1,356
   4,078
+   922
⎯⎯⎯
6,356

Be sure to line up the digits by place-value position.

TRY These

Add.

1. 8
  9
+ 2

2. 6
  7
+ 4

3. $27
   32
+ 61

4. 14
  96
+ 29

5. 8 + 5 + 9 + 1

6. 24 + 62 + 31 + 62

7. 82 + 59 + 64 + 71

Exercises

Add.

1. 21
    30
    85
    + 72

2. 33
    22
    67
    + 75

3. $39
    21
    75
    + 44

4. 143
    926
    47
    + 89

5. $13.89
    7.49
    82.21
    + 93.27

6. 324
    562
    + 731

7. 3,542
    7,256
    + 4,984

8. 1,498
    5,625
    + 216

9. $87.21
    93.75
    + 73.75

10. 331 + 405 + 791 + 652

11. 9,182 + 4,567 + 825

12. $8,460 + $7,208 + $5,032

Problem Solving

13. Brighton Middle School has 334 sixth graders, 464 seventh graders, and 310 eighth graders. How many students are at Brighton in all?

14. The Johnsonville Automobile Club sent out 1,736 books in June, 1,402 books in July, and 965 books in August. What is the total number of books sent during those three months?

5 Subtracting

Manuel knows how difficult it is to be sick for a long time, so he asked his classmates to donate 405 magazines and books to the hospital. If 237 of these were magazines, how many were books?

Subtract 237 from 405.

Estimate.
400 − 200 = 200

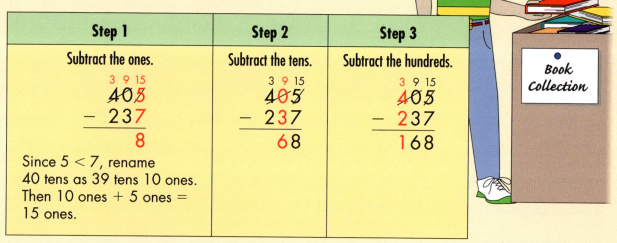

Step 1	Step 2	Step 3
Subtract the ones.	Subtract the tens.	Subtract the hundreds.
3 9 15 4̶0̶5̶ − 237 ——— 8	3 9 15 4̶0̶5̶ − 237 ——— 68	3 9 15 4̶0̶5̶ − 237 ——— 168
Since 5 < 7, rename 40 tens as 39 tens 10 ones. Then 10 ones + 5 ones = 15 ones.		

Manuel and his classmates donated 168 books.

Compared to the estimate, 200, the answer seems reasonable.

More Examples

A.
10
1̶0̶ 15
2̶1̶5̶
− 136
———
79

Check.
136
+ 79
———
215

B.
2 9 10
$3̶.0̶0̶
− 0.68
———
$2.32

C.
10
7 0̶ 15
8̶1̶5̶
− 98
———
717

Be careful. Some zeros are hidden.

TRY These

Subtract.

1. 2 9 16
    3̶0̶6̶   6 < 8
    − 128 Rename
    ———   30 tens.

2. 6 9 11
    7̶0̶1̶   1 < 9
    − 299 Rename
    ———   70 tens.

3. 915
    − 318
    ———

4. 630
    − 135
    ———

Exercises

Subtract.

1. 703 − 344
2. 848 − 246
3. 606 − 255
4. 876 − 652
5. 804 − 206

6. 905 − 330
7. 203 − 68
8. 340 − 27
9. 621 − 452
10. $2.67 − 1.29

11. $6.07 − 3.30
12. $8.01 − 3.52
13. 900 − 525
14. 621 − 342
15. 480 − 120

16. 600 − 327
17. 949 − 751
18. $205 − 166
19. $533 − 244
20. 308 − 116
21. 470 − 306
22. 510 − 77
23. 130 − 57

Problem Solving

24. There are 220 fifth graders. There are 27 fewer sixth graders. How many total students are in the fifth and sixth grades?

25. How much more did Arnie spend than Andy?

Arnie		Andy	
Apple	75¢	Apple	75¢
Orange	75¢	Orange	75¢
Grapes	80¢	Grapes	51¢
Pear	75¢	Pear	75¢

26. On one Saturday, 73,927 people attended a football game. The next week 68,412 people went to the game. What is the difference in attendance for the two games?

5 Subtracting

6 Multiplying by 1-Digit Numbers

The Fancy Fruit Orchard has 1,317 fruit trees. Suppose 5 bushels of fruit are harvested from each tree. To find the total number of bushels of fruit, multiply. Estimate. 5 × 1,000 = 5,000

Step 1	Step 2	Step 3
Multiply the ones.	Multiply the tens.	Multiply hundreds and thousands.
$\begin{array}{r} \overset{3}{1,317} \\ \times 5 \\ \hline 5 \end{array}$	$\begin{array}{r} \overset{3}{1,317} \\ \times 5 \\ \hline 85 \end{array}$	$\begin{array}{r} \overset{1}{1,}\overset{3}{3}17 \\ \times 5 \\ \hline 6,585 \end{array}$
5 × 7 ones = 35 ones Rename 35 ones as 3 tens + 5 ones.	5 × 1 ten = 5 tens 5 tens + 3 tens = 8 tens	5 × 3 hundreds = 15 hundreds Rename. 5 × 1 thousand = 5 thousands 5 thousands + 1 thousand = 6 thousands

There are 6,585 bushels of fruit.

Since 1,000 is much less than 1,317, 5,000 is a low estimate, so the answer makes sense.

More Examples

A. $\begin{array}{r} \overset{1}{23} \\ \times 4 \\ \hline 92 \end{array}$ **B.** $\begin{array}{r} \overset{4}{608} \\ \times 6 \\ \hline 3,648 \end{array}$ **C.** $\begin{array}{r} \overset{132}{\$127.52} \\ \times 4 \\ \hline \$510.08 \end{array}$ Multiply like whole numbers. Show dollars and cents in the answer.

TRY These

Multiply.

1. $\begin{array}{r} \$37 \\ \times 6 \\ \hline \end{array}$ 2. $\begin{array}{r} 232 \\ \times 2 \\ \hline \end{array}$ 3. $\begin{array}{r} 306 \\ \times 8 \\ \hline \end{array}$ 4. $\begin{array}{r} 9,324 \\ \times 6 \\ \hline \end{array}$

5. $\begin{array}{r} 1,256 \\ \times 4 \\ \hline \end{array}$ 6. $\begin{array}{r} \$80.07 \\ \times 6 \\ \hline \end{array}$ 7. $\begin{array}{r} \$10.60 \\ \times 9 \\ \hline \end{array}$ 8. $\begin{array}{r} 81,208 \\ \times 8 \\ \hline \end{array}$

Exercises

Multiply.

1. $0.83 × 5
2. 95 × 4
3. $4.05 × 8
4. 510 × 6
5. 623 × 3

6. 3,152 × 5
7. 5,670 × 7
8. 8,765 × 3
9. 4,412 × 6
10. 9,346 × 8

11. 86 × 5
12. 612 × 9
13. 7 × $4.52

14. 3 × 15,674
15. 9 × $49.35
16. 3 × $835.92

17. 48 × 5
18. 2 × 328
19. 9 × 218

20. What is 8,496 times 7?

21. What is the product of 3 and $17.50?

Problem Solving

22. Use the data on the previous page. If a bushel of apples is sold for $8.95, how much would the Fancy Fruit Orchard receive for the apples on one tree?

23. How much do 3 of these trucks weigh?

weight: 5,362 pounds

24. If popsicles cost $1.95 per dozen, what would be the cost for 8 dozen?

25. Debbie baked 24 cookies on Monday. If she bakes the same amount on Tuesday and Wednesday, how many cookies will she have baked?

7 Related Facts and Missing Factors

Koto places 4 photographs on each page of an album. How many pages are needed for 12 photographs?

▶ Divide to find the number of groups.

You can think of a related multiplication fact to help you divide.

■ × 4 = 12   **THINK** 4 × _?_ = 12

3 × 4 = 12, so 12 ÷ 4 = 3

Koto needs 3 album pages for the photographs.

More Examples

You can show division in different ways.

A. $4\overline{)12}^{\,3}$ quotient / divisor$)$dividend

B. $\dfrac{12}{4} = 3$

▶ Division can be shown as a fraction.

Division undoes multiplication. You can show this with fact families.

C. 42, 6, 7

6 × 7 = 42   42 ÷ 7 = 6
7 × 6 = 42   42 ÷ 6 = 7

D. 8, 1, 8

8 × 1 = 8   8 ÷ 1 = 8
1 × 8 = 8   8 ÷ 8 = 1

TRY These

Write each fact family.

1. 45, 9, 5
2. 5, 30, 6
3. 6, 1, 6

Copy and complete.

4. ■ × 8 = 40, so
   40 ÷ 8 = ■

5. ■ × 7 = 0, so
   0 ÷ 7 = ■

6. 9 × ■ = 27, so
   27 ÷ 9 = ■

7. 6 × ■ = 18, so
   18 ÷ 6 = ■

8. ■ × 5 = 20, so
   20 ÷ 5 = ■

9. 1 × ■ = 8, so
   8 ÷ 1 = ■

Exercises

Write each fact family.

1. 32, 8, 4
2. 6, 2, 12
3. 18, 3, 6
4. 3, 5, 15
5. 36, 9, 4
6. 4, 1, 4
7. 8, 7, 56
8. 7, 4, 28
9. 48, 8, 6
10. 24, 3, 8

Divide. Think of related multiplication facts.

11. $72 \div 9$
12. $9 \div 3$
13. $54 \div 6$
14. $25 \div 5$
15. $7 \div 7$
16. $8 \div 4$
17. $21 \div 3$
18. $10 \div 5$
19. $14 \div 7$
20. $36 \div 6$
21. $7\overline{)49}$
22. $1\overline{)5}$
23. $9\overline{)63}$
24. $6\overline{)24}$
25. $3\overline{)6}$
26. $8\overline{)16}$
27. $9\overline{)81}$
28. $8\overline{)64}$
29. $4\overline{)16}$
30. $5\overline{)35}$
31. $\frac{18}{9}$
32. $\frac{9}{9}$
33. $\frac{12}{3}$
34. $\frac{20}{4}$
35. $\frac{3}{3}$
36. $\frac{6}{3}$
37. $\frac{21}{7}$
38. $\frac{42}{6}$
39. $\frac{24}{4}$
40. $\frac{0}{5}$

41. What is the quotient when zero is divided by any number?
42. Name *four* division facts that have a quotient of 8.

PROBLEM Solving

43. Monica puts 9 stickers on each poster. If she has 36 stickers, how many posters does she use?

44. If water rushing over Niagara Falls washes away 3 inches of the ledge each year, in how many years will 12 inches be washed away?

45. Each box contains 8 crayons. How many boxes hold 24 crayons altogether?

8 Mental Math Computation: Division Patterns

Mr. Brown divided 40 baseball cards equally between his 2 children. How many cards did each child receive?

Each child received the same number of cards, so you can divide. Using division facts can help you divide greater numbers.

▶ Divide to find how many are in each group.

THINK $4 \div 2 = 2$

So, $40 \div 2 = 20$.   4 tens ÷ 2 = 2 tens   Each child received 20 cards.

More Examples

A. $12 \div 4 = 3$
    $120 \div 4 = 30$
    $1{,}200 \div 4 = 300$

B. $5\overline{)30}^{6}$ $5\overline{)300}^{60}$ $5\overline{)3{,}000}^{600}$

C. $\dfrac{56}{8} = 7$ $\dfrac{560}{8} = 70$ $\dfrac{5{,}600}{8} = 700$

TRY These

Write each quotient.

1. $15 \div 3$ 2. $24 \div 8$ 3. $45 \div 9$ 4. $36 \div 6$
    $150 \div 3$ $240 \div 8$ $450 \div 9$ $360 \div 6$
    $1{,}500 \div 3$ $2{,}400 \div 8$ $4{,}500 \div 9$ $3{,}600 \div 6$

5. $8\overline{)40}$ $8\overline{)400}$ $8\overline{)4{,}000}$ 6. $\dfrac{54}{9}$ $\dfrac{540}{9}$ $\dfrac{5{,}400}{9}$

Exercises

Divide.

1. 160 ÷ 4   16 ÷ 4 = 4
2. 210 ÷ 3   21 ÷ 3 = 7
3. 350 ÷ 5
4. 420 ÷ 7
5. 270 ÷ 3
6. 540 ÷ 9
7. $\dfrac{360}{6}$
8. $\dfrac{180}{2}$
9. $\dfrac{300}{5}$
10. $\dfrac{280}{4}$
11. $\dfrac{150}{3}$
12. $\dfrac{120}{6}$
13. $\dfrac{490}{7}$
14. $\dfrac{640}{8}$
15. $\dfrac{630}{9}$
16. $\dfrac{100}{5}$
17. 5)250
18. 4)320
19. 8)640
20. 6)420
21. 9)630
22. 4)3,200
23. 9)7,200
24. 7)7,000
25. 6)3,000
26. 3)1,800
27. 9)36,000
28. 8)80,000
29. 2)10,000
30. 3)18,000
31. What is 240,000 divided by 4?
32. What is 300,000 divided by 5?

PROBLEM Solving

33. If there are 40 cards in 4 sets of baseball cards, how many cards are in 1 set?
34. Light bulbs are sold 4 to a package. How many packages are needed for 3,200 light bulbs?
35. At a banquet, 120 people were seated 4 to a table. How many tables were used?
36. Walter drove his car 180 miles on 9 gallons of gas. How many miles did he get to the gallon on average?
37. In 5 months Hector saved $350. He saved the same amount each month. How much did he save each month?

8 Mental Math Computation: Division Patterns

9 Division with Remainders

Peta has 39 silk flowers. She plans to use 5 flowers in each corsage. How many corsages can she make? How many flowers will be left over?

Since Peta uses the same number of flowers in each corsage, divide 39 by 5.

Step 1	Step 2	Step 3
Divide. $\quad$ 7 $\quad$ 5)39	Multiply. $\quad$ 7 $\quad$ 5)39 $\quad$ −35 $\quad$ 5 × 7	Subtract and compare. $\quad$ 7 R4 $\quad$ 5)39 $\quad$ −35 $\quad\quad$ 4 ← remainder
THINK $5 \times 8 = 40$ The quotient is too great! Try 5×7.	35 flowers are used to make 7 corsages.	Compare to be sure the remainder is less than the divisor.

Check:
$\quad$ 7 quotient
$\times$ 5 divisor
$\quad$ 35
$+$ 4 remainder
$\quad$ 39 dividend

Peta can make 7 corsages.
She will have 4 flowers left over.

More Examples

A. 3)11 −9 = 3 R2, remainder 2

B. 6)52 −48 = 8 R4, remainder 4

C. 4)24 −24 = 6, remainder 0

D. 8)76 −72 = 9 R4, remainder 4

TRY These

Divide.

1. 4)9
2. 5)22
3. 6)15
4. 8)25

5. 6)33
6. 9)17
7. 5)40
8. 7)37

Exercises

Divide.

1. 3)26
2. 6)42
3. 4)19
4. 5)48
5. 9)64
6. 4)35
7. 9)30
8. 2)11
9. 7)25
10. 8)55
11. 6)57
12. 4)39
13. 8)64
14. 7)27
15. 7)44
16. 8)62
17. 9)35
18. 2)19
19. 6)58
20. 7)50
21. 29 ÷ 3
22. 86 ÷ 9
23. 15 ÷ 2
24. 28 ÷ 5

Problem Solving

25. Holly has 36 coins. How many rows can she make if she puts 6 coins in each row? 9 coins in a row? 3 coins in a row?

26. In an hour, 5 people delivered 45 telephone directories altogether. Each person delivered the same number of directories. How many directories did each person deliver?

9 Division with Remainders

10 Dividing 2-Digit Numbers

Charlene collected 87 buttons. If she puts an equal number of buttons on 4 cards, how many buttons will be on each card?

Step 1	Step 2	Step 3
87 = 8 tens 7 ones	Divide the tens.	Divide the ones.
$4\overline{)87}$	$\begin{array}{r}2\\4\overline{)8\,7}\\-8\\\hline 0\end{array}$ 8 tens ÷ 4 = 2 tens; 4 × 2 tens = 8 tens; Subtract and compare. 0 < 4	$\begin{array}{r}2\,1\ \text{R3}\\4\overline{)8\,7}\\-8\\\hline 0\,7\\-4\\\hline 3\end{array}$ Divide. 7 ÷ 4; Multiply. 4 × 1 = 4; Subtract and compare. 3 < 4

Divide to find how many in each group.

There will be 21 buttons on each card. There will be 3 buttons left.

More Examples

A. $\begin{array}{r}32\\2\overline{)64}\\-6\\\hline 04\\-4\\\hline 0\end{array}$

B. $\begin{array}{r}12\ \text{R5}\\7\overline{)89}\\-7\\\hline 19\\-14\\\hline 5\end{array}$

Check:
$\begin{array}{r}12\\\times\ 7\\\hline 84\\+\ \ 5\\\hline 89\end{array}$

TRY These

Divide.

1. $2\overline{)64}$ 2. $4\overline{)48}$ 3. $8\overline{)92}$ 4. $4\overline{)89}$ 5. $4\overline{)47}$

Exercises

Divide.

1. 6)67
2. 2)43
3. 3)95
4. 4)86
5. 7)77

6. 5)92
7. 3)57
8. 6)72
9. 3)85
10. 4)78

11. 6)98
12. 5)75
13. 2)64
14. 3)97
15. 4)65

16. 5)74
17. 9)98
18. 7)82
19. 4)72
20. 4)53

21. 68 ÷ 3
22. 75 ÷ 2
23. 80 ÷ 5
24. 84 ÷ 7
25. 96 ÷ 8

Problem Solving

Use the chart for problems 26–27.

26. Cars will be used to get to the picnic. If each car holds 5 students, how many cars are needed?

27. Eight families shared the cost of the food evenly. How much did each family pay?

28. In a bicycle race Jane rode her bicycle 92 kilometers in 4 hours. How many kilometers did she average per hour?

29. James is packaging books. He is placing 4 books in a box. How many boxes will he need for 64 books?

Grade 5 Picnic
Number of students: 43
Amount spent for food: $96

10 Dividing 2-Digit Numbers

11 Dividing 3-Digit Numbers

One Girl Scout project is to make door decorations using pinecones. Mrs. Lopez needs 632 pinecones for her scout troops. The cones are sold in packages of 5. How many packages should she buy?

Step 1	Step 2	Step 3
Divide the hundreds.	**Divide the tens.**	**Divide the ones.**
$\begin{array}{r} 1 \\ 5{\overline{\smash{)}632}} \\ \underline{-5} \\ 1 \end{array}$ 6 hundreds ÷ 5 5 × 1 hundred Subtract and compare. 1 < 5	$\begin{array}{r} 12 \\ 5{\overline{\smash{)}632}} \\ \underline{-5}\downarrow \\ 13 \\ \underline{-10} \\ 3 \end{array}$ 13 tens ÷ 5 5 × 2 tens Subtract and compare. 3 < 5	$\begin{array}{r} 126 \text{ R2} \\ 5{\overline{\smash{)}632}} \\ \underline{-5} \\ 13 \\ \underline{-10}\downarrow \\ 32 \\ \underline{-30} \\ 2 \end{array}$ 32 ÷ 5 5 × 6 2 < 5
▶ Division is used to find how many groups.		

Mrs. Lopez needs 126 full packages and 2 more pinecones. She should buy 127 packages. Why?

More Examples

A. $\begin{array}{r} 157 \\ 4{\overline{\smash{)}628}} \\ \underline{-4}\downarrow \\ 22 \\ \underline{-20}\downarrow \\ 28 \\ \underline{-28} \\ 0 \end{array}$

B. $3{\overline{\smash{)}\$5.04}} \longrightarrow \begin{array}{r} 168 \\ 3{\overline{\smash{)}5.04}} \\ \underline{-3} \\ 20 \\ \underline{-18} \\ 24 \\ \underline{-24} \\ 0 \end{array} \longrightarrow \begin{array}{r} \$1.68 \\ 3{\overline{\smash{)}\$5.04}} \end{array}$

▶ Divide money like whole numbers. Show dollars and cents in the quotient.

TRY These

Divide.

1. $2{\overline{\smash{)}653}}$ 2. $5{\overline{\smash{)}\$8.40}}$ 3. $3{\overline{\smash{)}971}}$ 4. $6{\overline{\smash{)}714}}$ 5. $7{\overline{\smash{)}977}}$

Exercises

Divide.

1. 5)732
2. 4)642
3. 2)$5.10
4. 7)934
5. 4)894
6. 8)906
7. 3)$6.78
8. 5)964
9. 6)857
10. 2)915
11. 4)513
12. 6)$8.28
13. 2)550
14. 8)987
15. 7)970
16. 5)709
17. 3)345
18. 4)647
19. 6)$7.62
20. 2)311
21. $460 ÷ 4
22. 935 ÷ 3
23. 789 ÷ 6
24. 536 ÷ 2
25. 500 ÷ 3
26. 812 ÷ 5
27. 646 ÷ 4
28. 847 ÷ 7
29. 9,713 ÷ 8
30. $91.84 ÷ 7
31. 5,490 ÷ 3

Problem Solving

32. **Use the data on the previous page.** Mrs. Lopez gives each scout 8 pinecones for the project. How many scouts are there?

33. Bill's market received a total of 402 oranges. They were packed in 3 crates. Find the number of oranges in each crate.

34. How many milliliters are in each test tube?

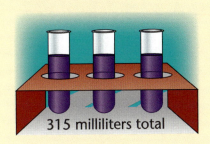
315 milliliters total

11 Dividing 3-Digit Numbers

12 More Dividing

There are 220 feet of ribbon on a spool. How many pieces 1 yard long can be cut? Since 1 yard = 3 feet, divide 220 by 3.

Estimate.
210 is close to 220.
3 × 7 tens = 21 tens
210 ÷ 3 = 70

Step 1	Step 2	Step 3
Divide the hundreds.	**Divide the tens.**	**Divide the ones.**
3)220	7 3)220 −21 ___ 1	73 R1 3)220 −21↓ ___ 10 − 9 ___ 1
Are there enough hundreds to divide by 3? No. So, you can rename 2 hundreds 2 tens as 22 tens.	Divide. Multiply. Subtract. Compare.	Rewrite ones. Divide. Multiply. Subtract. Compare. 1 < 3

You can cut 73 pieces of ribbon each 1 yard long. The remainder shows that 1 foot of ribbon will be left.

Compared to the estimate of 70, the answer 73 is reasonable.

More Examples

A. 98 R3
    4)395
    −36↓
    ___
     35
    − 32
    ___
      3

   98
   × 4
   ___
   392 + 3 = 395

   Division can be checked by multiplication.

B. 5)$4.25 →
     85
   5)425
   −40↓
   ___
    25
   − 25
   ___
     0

   → $0.85
   5)$4.25

   $ 0.85
   ×   5
   _____
   $4.25

   Divide as usual. Show dollars and cents in the quotient.

TRY These

Divide.

1. 8)437
2. 2)$186
3. 5)355
4. 8)729
5. 7)458

6. 5)125
7. 9)357
8. 6)$5.58
9. 9)643
10. 7)236

Exercises

Divide.

1. 6)495
2. 7)653
3. 5)426
4. 3)$8.55
5. 7)506
6. 3)127
7. 6)472
8. 5)$9.60
9. 9)728
10. 6)432
11. 4)857
12. 5)364
13. 7)515
14. 3)742
15. 6)284
16. 139 ÷ 2
17. 428 ÷ 5
18. $476 ÷ 7

19. What is 539 divided by 8?

Problem Solving

20. Walt has 135 baseball cards in 9 boxes. If each box holds the same number of cards, how many are in each box?

21. The distance from Rome to Paris is 1,098 kilometers. If a plane takes 3 hours to make the trip, how many kilometers did it average each hour?

13 Fractions Show Parts of a Whole and Parts of a Group

How much of Josh's ice cream is strawberry? Three of the four scoops, or three-fourths of the ice cream on the cone is strawberry.

Three-fourths can be written as the fraction $\frac{3}{4}$.

$\frac{3}{4}$ ← numerator — number of parts you are talking about
← denominator — number of congruent parts in the whole

3 parts colored out of 4 congruent parts in the whole

Another Example

Fractions can also show parts of a set.

There are 16 ounces in a pound. Make a model to show $\frac{3}{8}$ of a pound.

1. Use counters to show the whole set.

2. Separate the set into 8 equal groups to show eighths. (●● is $\frac{1}{8}$.)

3. Find the groups of counters that show 3 of the eighths.

   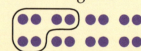

$\frac{3}{8}$ ← number of groups you are talking about
← number of congruent groups in the whole set

There are 6 ounces in $\frac{3}{8}$ of a pound.

TRY These

Write a fraction to name the colored part of each whole.

1.
2.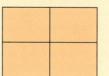
3.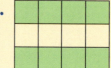
4.

Exercises

Draw two different models to show each fraction.

1. $\frac{5}{6}$
2. three-tenths
3. $\frac{3}{9}$
4. seven-eighths

Draw a model to show each set. Then use the model to complete each statement.

5. A year has 12 months. $\frac{2}{3}$ of a year has ■ months.
6. An hour has 60 minutes. $\frac{5}{6}$ of an hour has ■ minutes.
7. A pizza is cut into eighths. $\frac{3}{4}$ of the pizza equals ■ slices.
8. A sheet cake is cut into twentieths. $\frac{4}{5}$ of the cake equals ■ pieces.

Copy and complete. Use the models.

9. $\frac{1}{3}$ of 9 = ■
10. $\frac{1}{2}$ of 4 = ■
11. $\frac{1}{5}$ of 10 = ■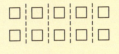
12. $\frac{1}{4}$ of 8 = ■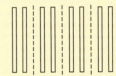

13. $\frac{1}{■}$ of 12 = 4
14. $\frac{2}{3}$ of ■ = 8
15. $\frac{3}{8}$ of 16 = ■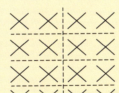
★ 16. $\frac{2}{■}$ of 15 = 6

Problem Solving

17. Pat makes a cake with 5 layers. The top 2 layers are banana nut. What fraction of the cake is banana nut?

18. Barbara has 10 rosebushes and 7 have red roses. What fraction of the bushes have red roses?

19. Mimi gets change for $1 in dimes. She spends $\frac{2}{5}$ of her dollar. How much does she spend?

20. Mel cuts the cake into 6 equal parts so that 6 people each get an equal share. What fraction of the cake is each slice?

13 Fractions Show Parts of a Whole and Parts of a Group

Diagnostic Skills Posttest

Write the value of the underlined digit. (Section 1)

1. 1<u>9</u>,806
2. 940,8<u>1</u>6
3. <u>8</u>0,321
4. <u>8</u>18,376

Write in standard form. (Section 1)

5. twenty-nine thousand, thirty-six
6. forty-five thousand, three hundred two
7. 800,000 + 50,000 + 600
8. 10,000 + 3,000 + 800 + 60

Round to the underlined place-value position. (Section 2)

9. 3,<u>8</u>12
10. <u>5</u>,915
11. 9<u>0</u>5
12. <u>8</u>20

Add. (Section 3)

13. 2,621 + 1,659
14. 6,314 + 5,918
15. $76.26 + $3.28
16. $443.13 + $77.67

Add. (Section 4)

17. 42 + 19 + 28
18. 37 + 87 + 13
19. 412 + 546 + 817
20. 5,923 + 876 + 24

Subtract. (Section 5)

21. 301 − 17
22. 651 − 432
23. 900 − 32
24. $305.00 − 57.24

Multiply. (Section 6)

25. 76 × 3
26. 804 × 9
27. 4,609 × 4
28. $23.53 × 5

Write the fact family for each set of numbers. (Section 7)

29. 6, 9, 54
30. 4, 16, 4
31. 8, 8, 1
32. 40, 8, 5

Divide. (Section 8)

33. 9)540   34. 6)3,000   35. 8)720   36. 8)240

Divide. (Section 9)

37. 5)41   38. 8)33   39. 9)19   40. 7)22

Divide. (Section 10)

41. 7)85   42. 2)39   43. 4)98   44. 6)94

Divide. (Section 11)

45. 6)876   46. 6)$8.10   47. 3)965   48. 5)6,320

Divide. (Section 12)

49. 9)508   50. 4)128   51. 6)$4.14   52. 4)334

Write a fraction to name the colored part of each whole or each set. (Section 13)

53. 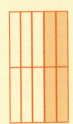   54.

Complete. (Section 13)

55. $\frac{2}{3}$ of 15 = ■   56. $\frac{1}{4}$ of 24 = ■

Diagnostic Skills Posttest   xliii

CHAPTER 1

Numbers and Place Value

William Braun
Minnesota

1.1 Renaming Whole Numbers

Objective: to rename whole numbers

A census (count) of people living in the United States has been taken every ten years since 1790. There were 210 ten-year periods from 1790 to 2000.

Two hundred ten can be renamed as 210 ones or 21 tens.

There are 21 10-year periods from 1790 to 2000.

Use place-value positions to rename numbers.

210 ones
or
21 tens

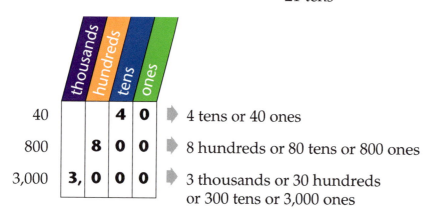

40 → 4 tens or 40 ones

800 → 8 hundreds or 80 tens or 800 ones

3,000 → 3 thousands or 30 hundreds or 300 tens or 3,000 ones

Examples

A. 3,945 = 39 hundreds 45 ones

B. $9,400 = 94 one-hundred dollar bills

C. 723 tens 5 ones = 7,235

TRY These

Copy and complete.

1. 20 = ■ tens
2. 80 = ■ tens
3. 600 = ■ tens
4. 580 = ■ ones
5. 3,600 = ■ hundreds
6. 1,400 = ■ tens

Exercises

Copy and complete.

1. 690 = ■ tens
2. 7,400 = ■ hundreds
3. 8,000 = ■ hundreds
4. $780 = ■ ten-dollar bills
5. 605 = ■ tens ■ ones
6. 4,693 = ■ tens ■ ones
7. 9,137 = ■ hundreds ■ ones
★ 8. 86,000 = ■ hundreds
★ 9. 370,000 = ■ thousands

Write in standard form.

10. 50 hundreds
11. 75 tens
12. 76 tens 3 ones
13. 625 ten-dollar bills
14. 384 tens 9 ones
15. 42 thousands 360 ones

Problem Solving

16. Which of the following is *not* the same as 5,489?

    a. 5 thousands 47 tens 19 ones
    b. 54 hundreds 98 ones
    c. 4 thousands 148 tens 9 ones

★ 17. Write 63 thousands 43 hundreds 23 ones in standard form. (*Hint:* Use a place-value chart.)

★ 18. Pretend you just won $100,000 in 10-dollar bills. How many $10 bills did you win?

Mind Builder

History

The abacus is an early form of today's calculators and computers. In this version of the Chinese abacus, the beads in the upper section each represent 5 units; the beads in the lower section each represent 1 unit. The beads are moved to the center bar to show 2,783.

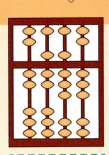

▶ Does the abacus show place value?

Draw an abacus to show each number.

1. 46
2. 172
3. 2,091

1.1 Renaming Whole Numbers

1.2 Millions and Billions

Objective: to apply knowledge of place value through billions

In a recent year the estimated population of the world was about 6,500,000,000.

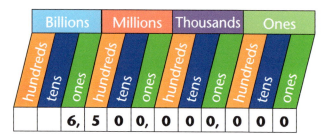

The number that represents the population is written below in three forms.

Standard Form   6,500,000,000

Expanded Form   6,000,000,000 + 500,000,000

Words   six billion, five hundred million

Each digit has a value depending on its place.

Another Example

Find the value of the underlined digit in 567,341,456.

The 7 is in the millions place so it has a value of 7,000,000.

Write the value of the underlined digit.

1. 3,0<u>9</u>5,230
2. <u>2</u>,346,214,300
3. <u>4</u>32,671,030,438
4. <u>8</u>30,500,000
5. 6<u>7</u>0,432,192
6. 172,<u>5</u>36,148

8,00 000 000 + 30 000 000 + 5 00000

Exercises

Write in standard form, expanded form, and words.

1. seven million
2. 5,236,000
3. 400,000,000 + 80,000,000 + 500,000 + 70,000 + 1,000
4. five hundred twenty-four billion, six hundred thousand
5. 72,620,005
6. 5,000,000 + 300,000 + 70,000 + 5,000 + 300   5,375,300
7. 831,724
8. one hundred twenty-three million, three thousand, four hundred seventy-two
9. 400,000 + 60,000 + 2,000 + 400 + 90 + 7   123,003,472
10. 305,582,492,116

PROBLEM Solving

11. What number is ten thousand less than 765,352,011?   535,2011
12. Use the digits 1, 2, 4, 6, 8, and 9 to make the greatest and least possible numbers.

Constructed Response

13. In Calvert's warehouse there are 7 pallets with 10,000 pennants each, 5 cartons with 1,000 pennants each, and 9 boxes with 100 pennants each. Write the total number of pennants in standard form, expanded form, and words. Explain the differences between the three forms.

TEST Prep

14. Which of the following does not equal 3,409?
    a. 34 hundreds 9 ones
    b. 340 tens 9 ones
    c. 34 thousands 9 ones
    d. 3,409 ones

1.3 Comparing and Ordering Whole Numbers

Objective: to compare and order whole numbers

The table shows the greatest known depths of the world's four largest oceans.

Ocean Depths	
Ocean	Greatest Known Depth (feet)
Atlantic	27,493
Arctic	18,050
Indian	24,442
Pacific	36,198

To Compare

Which is deeper, the Atlantic Ocean or the Indian Ocean?

To compare the depths of the Atlantic and Indian oceans first line them up vertically. Keep the place values in line. Remember > means greater than and < means less than.

Atlantic Ocean   27,493
Indian Ocean   24,442

> Start with the greatest place value. Compare the digits. You can see that 7 is greater than 4.

The Atlantic Ocean is deeper because 27,493 > 24,442.

To Order

List the ocean depths in order from greatest to least and least to greatest. To order a list of numbers, begin by writing the numbers in a vertical list, keeping the place values in line.

27,493
24,442
36,198
18,050

Start with the greatest place-value position. Compare the digits.

You can see that 18,050 is the least because the 1 is in the ten thousands place and it is the least number in that place.

You can see that 36,198 is the greatest because 3 is in the ten thousands place and it is the greatest number in that place.

To compare 27,493 and 24,442 look at the digit in the thousands place. Since 4 < 7, the number 24,442 is less than 27,493.

In order from greatest to least, the depths are:

36,198 (Pacific)   27,493 (Atlantic)   24,442 (Indian)   18,050 (Arctic)

In order from least to greatest, the depths are:

18,050 (Arctic)   24,442 (Indian)   27,493 (Atlantic)   36,198 (Pacific)

TRY These

Write using the symbols <, >, or =.

1. 1,234 is equal to 1,234.
2. 8,961 is less than 8,962.
3. 2,681 is greater than 1,998.
4. 4,005 is less than 4,099.

Compare using <, >, or =.

5. 4,709 ● 4,710
6. 39,499 ● 34,999
7. 8,309 ● 8,390

Exercises

Compare using <, >, or =.

1. 74,293 ● 74,293
2. 13,089 ● 13,809
3. 486,926 ● 478,926
4. 924,681 ● 924,618
5. 5,313,313 ● 5,331,313
6. 3,256,729 ● 3,729,256
7. 365,291,074 ● 365,291,074
8. 62,850,000 ● 62,580,000

Order from least to greatest.

9. 1,020   1,200   1,120   2,010
10. 2,063   2,060   2,058
11. 41,263   40,263   39,263
12. 40,007   47,000   40,700

1.3 Comparing and Ordering Whole Numbers

Order from greatest to least.

13. 82,541   28,541   82,145

14. 330,303,330   330,330,303   303,330,033

Write a number for each missing digit to make a true inequality.

15. 6,409 > 6,■09

16. 48,863 > 48,■63

Compare.

17. Which is greater, 300 thousand or 30 million?

18. Which is least, 8 ten thousands, 2 millions, or 50,000?   *50,000*   *8000*

Use the chart to answer questions 19–22.

19. Which state has the least area?

20. Which state has an area more than 8,000 square miles and less than 9,000 square miles?

21. List the states in order by size. Start with the smallest state.

22. Maryland has an area of 12,407 square miles. If Maryland were added to your ordered list, between which two states would it fall?

Areas of Some Eastern States	
State	Area (in square miles)
Connecticut	5,009
Delaware	2,057
Maine	33,215
Massachusetts	8,257
New Hampshire	9,304
New York	49,576
Rhode Island	1,214

Write in standard form.

23. 58 tens 7 ones

24. 16 thousands 43 ones

25. 8 thousands 65 tens

26. 382 tens

8   1.3 Comparing and Ordering Whole Numbers

Cumulative Review

Make each number greater by the given amount.

1. 69,638 + 1,000
2. 37,846 + 100
3. 78,351 + 10,000
4. 24,958 + 100,000

Write the value of each underlined digit.

5. <u>8</u>,630
6. 9<u>9</u>9
7. 3<u>2</u>,621
8. 40,4<u>5</u>9
9. <u>3</u>7,291
10. 876,54<u>3</u>
11. 100,1<u>1</u>0
12. <u>4</u>56,709

Write in standard form, expanded form, and words.

13. six hundred seventy-eight
14. nine thousand
15. three hundred twenty thousand
16. eighty-two thousand, fifty
17. one hundred one thousand, ten
18. seven hundred thousand, seven
19. 16
20. 365
21. 4,270
22. 98,706
23. 200,300
24. 40 + 7
25. 200 + 60 + 8
26. 900 + 8
27. 5,000 + 200 + 9
28. 30,000 + 1,000 + 3

Write the number and operation you would use to get the second number.

29. 67,914 ⟶ 68,014
30. 138,472 ⟶ 38,472
31. 2,375,900 ⟶ 2,376,000
32. 90,350 ⟶ 89,350

Copy and complete.

33. 50 = ■ tens
34. 360 = ■ tens
35. 2,000 = ■ tens
36. 600 = ■ hundreds
37. 2,500 = ■ hundreds
38. 36,100 = ■ hundreds
39. 200,000 = ■ hundreds

Solve.

40. You want to buy a book for $8.25. You have saved $4.58. How much money do you still need?
41. Jan has 2 more sisters than Pepe. Jan has 5 sisters. How many sisters does Pepe have?

1.4 Rounding Whole Numbers

Objective: to round whole numbers

In a recent year, Americans owned 73,948,306 dogs. You can round this number to the nearest million without using a number line.

Look at the digit at the right of the place being rounded. Then round as follows.

Round up if the digit at the right is 5, 6, 7, 8, or 9.

To the nearest million, 73,948,306 rounds to 74,000,000.

The underlined digit stays the same if the digit at the right is 0, 1, 2, 3, or 4.

More Examples

A. 23,586 rounded to the nearest hundred is 23,600.

B. 574,298 rounded to the nearest hundred thousand is 600,000.

C. 32,599 rounded to the nearest ten thousand is 30,000.

D. 9,896,863 rounded to the nearest million is 10,000,000.

TRY These

Choose the correct answer for rounding to the nearest thousand.

1. 4,263   a. 4,000   b. 5,000
2. 36,522   a. 36,000   b. 37,000
3. 727,200   a. 727,000   b. 728,000
4. 88,099   a. 88,000   b. 89,000

Round to the nearest ten thousand.

5. 86,022   6. 41,949   7. 260,260
8. 107,291   9. 462,298   10. 91,910

Exercises

Choose the correct answer for rounding to the underlined place-value position.

1. 3_8_2   a. 380   b. 390
2. 12,3_6_5   a. 12,360   b. 12,370
3. 7,_6_45   a. 7,600   b. 7,700
4. 4_9_5,621   a. 490,000   b. 500,000
5. _9_69   a. 900   b. 1,000
6. 1,9_9_8   a. 1,990   b. 2,000

Round to the underlined place-value position.

7. 2_5_7
8. 4,0_8_1
9. _8_40
10. 5,0_2_9
11. _3_20,900
12. 4_6_,526,000
13. 43,7_2_8
14. 1_2_6,340,015
15. 37,_0_59
16. _9_81,235,628
17. 1_5_0,999
18. 2,_4_36,256

PROBLEM Solving

19. Micah has eaten 1,936 calories so far today. What is this number rounded to the nearest hundred?

★ 20. What is the greatest number that rounds to 43,800? the least number?

MIND Builder

Logical Reasoning

The people of Squareland write  to mean 47. They write to mean 293.

What do the following Squareland numerals mean?

1.
2.
3.
4.

How would you write the following numbers if you lived in Squareland?

5. 72   6. 364   7. 308   8. 400   9. 4,321   10. 9,030

1.5 Problem-Solving Strategy: Using the Four-Step Plan

Objective: to solve problems using the four-step plan

In the United States, a president serves a four-year term. In 1953, Dwight D. Eisenhower became president and went on to serve two consecutive terms. What was his last year in office?

You can use a four-step plan to help you solve problems.

1. READ
Read the problem carefully. Ask yourself questions like these.
a. "What am I trying to find?"
   the last year Eisenhower was in office
b. "Do I understand all the words?"
   Term means length of time. Consecutive means one follows the other.
c. "What facts do I know?"
   A term is 4 years.
   Eisenhower served two consecutive terms.
   He became president in 1953.

2. PLAN
See how the facts relate to each other. Decide how to solve the problem. Choose a strategy to help you.
Eisenhower became president in 1953. He served two terms. You can add to find his last year in office.

Write an equation to help you solve. Let n represent the unknown number.

$$n = 1953 + 8$$

n — last year in office
1953 — first year in office
8 — number of years served

3. SOLVE
Compute, or work through your plan.

1953 + 8 = 1961

Eisenhower's last year in office was 1961.
He was in office from January 1953 to January 1961.

4. CHECK
Reread the problem. Make sure you have answered the question. Ask yourself, "Does the answer make sense?"
Yes, because two terms last 8 years and 1953 + 8 = 1961.

TRY These

1. Dwight D. Eisenhower was born on October 14, 1890. How old was he when he became president in January of 1953?

2. Two of Dwight D. Eisenhower's brothers were older than he was and three were younger. Another brother died as an infant. How many sons did his parents have?

Solve

1. Bill is packaging pencils. He puts 24 red pencils and half as many yellow pencils in each box. How many pencils does he put in each box?

2. Martha read 214 books this year and 86 books last year. How many more books did she read this year?

3. There were 206 adults and children at the picnic. Of these, 69 were adults. How many children were at the picnic?

4. The library had 226 books. The librarian bought 58 new books for the library. How many books are in the library now?

5. Ginny was 47 inches tall last year. She grew 4 inches this year. How tall is she now?

6. Liz has $0.95. Does she have enough to buy a pencil that costs $0.55 and an eraser that costs $0.38?

MID-CHAPTER Review

Write in standard form, expanded form, and words.

1. 8,000,405
2. six hundred four thousand, twelve

Write the value of the underlined digit.

3. 2_7_,602
4. _5_03,721
5. 8_1_9,613
6. 295,_8_64

Round to the underlined place-value position.

7. 46,_2_19
8. 3_7_6,398
9. 5_8_9,621
10. 80_0_,999

1.5 Problem-Solving Strategy: Using the Four-Step Plan   13

1.6 Decimal Place Value

Objective: to understand decimal place value

In the women's gymnastic events, a gymnast finished fifth in the vault with 19.774 points.

The last digit is a 4.
Its place-value position is thousandths.
Its value is four thousandths or 0.004.

What numbers do the other digits name?

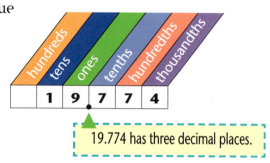

19.774 has three decimal places.

In words, 19.774 is "nineteen and seven hundred seventy-four thousandths." To name a decimal in words, first read the whole number part. Then read the decimal point as *and*. Finally read the decimal part and name the smallest place value of the number.

Another Example

You can use base ten blocks to build decimal models.

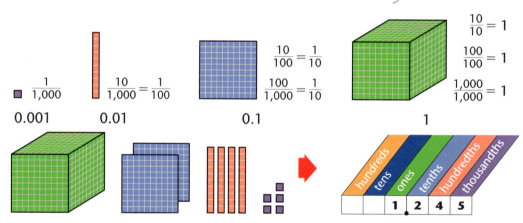

On a number line 1.245 comes between 1.24 and 1.26.

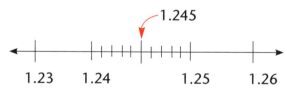

TRY These

Make a model of each number. Then write the value of each underlined digit in words.

1. 0.<u>7</u>
2. <u>4</u>.1
3. 0.0<u>5</u>
4. 0.1<u>5</u>
5. 0.04<u>2</u>

Exercises

Copy each number. Circle the digit in the tenths place-value position.

1. 37.365
2. 259.29
3. 0.673
4. 325.901

Copy each number. Circle the digit in the hundredths place-value position.

5. 7.29
6. 106.348
7. 29.21
8. 30.075

Copy each number. Circle the digit in the thousandths place-value position.

9. 24.389
10. 6.138
11. 52.230
12. 317.048

Write the value of each underlined digit in words.

13. 0.20_8_
14. 3.4_2_6
15. 31.0_1_
16. 3_6_2.7
17. _4_.002
18. 94.0_4_
19. 0._3_00
20. 1.30_9_
21. _1_6.48
22. 2.0_7_0

Problem Solving

23. Phoebe Mills received nineteen and eight hundred thirty-seven thousandths points for women's balance beam. Write the number as a decimal. 19.837,000

24. Derrick Adkins of the United States won the 400-meter hurdles with a time of forty-seven and fifty-four hundredths seconds. Write out this decimal.

25. The 400-meter relay was won by Canada with a time of 37.69 seconds. Write this number in words.

26. Mike Marsh of the United States won the 200-meter run with a time of $200\frac{1}{1,000}$ seconds. Write his time as a decimal.

Mind Builder

Attic Numerals

The early Greeks used a system of numeration called **Attic numerals**. These numerals are formed using the symbols I, △, H, and X for 1, 10, 100, and 1,000. A special symbol for 5, Γ, is used alone or with △, H, or X.

2,857 = XX ⌐△⌐ H H H ⌐Γ⌐ Γ II

Write 5,508 using Attic Greek numerals.

1.6 Decimal Place Value

1.7 Comparing and Ordering Decimals

Objective: to compare and order decimals

The winning time for both Hasely Crawford and Allan Wells was over 10 seconds. To find out who ran the 100-meter dash faster, compare 10.06 and 10.25.

Men's 100-Meter Dash		
Year	Winner	Time (seconds)
1976	Crawford	10.06
1980	Wells	10.25
1984	Lewis	9.99
1988	Lewis	9.92

THINK 10.06 and 10.25
The whole number parts are the same.
6 hundredths < 25 hundredths
10.06 < 10.25

Since Crawford's time was less than Wells's time, Crawford ran faster than Wells.

Examples

A. Compare 0.8 and 0.80.

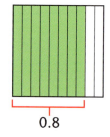
0.8

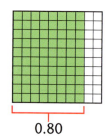
0.80

You can attach, or place, zeros to the right of a decimal without changing its value? Why?

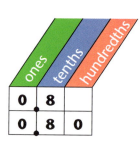

Equivalent fractions: $\frac{8}{10} = \frac{80}{100}$

Equivalent decimals: 0.8 = 0.80

Since the colored part of each square is the same, 0.8 is equal to 0.80.

B. Compare 9.84 and 9.96.
Compare the digits beginning at the greatest place-value position.

9.84
↑ ↑   9 = 9
↓ ↓   8 < 9
9.96

So, 9.84 < 9.96.

C. Order from least to greatest.
Line up by place value.

9.99
9.92
9.96
9.84

> The ones place all have a 9. Look to the next place. The 8 in the tenths place is less then 9, so 9.84 is the least. Compare the other numbers using the digit in the hundredths place.

In order from least to greatest the numbers are: 9.84, 9.92, 9.96, 9.99.

TRY These

Compare using <, >, or =. Explain your answer.

1. 0.6 ● 0.5
2. 0.18 ● 0.81
3. 2.40 ● 2.4
4. 9.90 ● 9.99

Exercises

Compare using <, >, or =.

1. 0.3 ● 0.6
2. 0.21 ● 0.19
3. 7.3 ● 7.2
4. 4.35 ● 4.53
5. 7.4 ● 7.400
6. 2.31 ● 2.30
7. 0.015 ● 0.201
8. 0.008 ● 0.010

Find a missing digit that will make the equation or inequality true.

9. 5.43 = 5.43■
10. 4.56 < 4.■6
11. 8.93 > 8.9■
12. 6.040 > 6.03■

Order from least to greatest.

13. 3.8   3   3.9   4.0
14. 8.8   8.08   8.808
15. 0.705   5.07   0.75   5

PROBLEM Solving

16. Copy the grid.
    Write one of the decimals shown in each square of the grid.
    The decimals must be in order from least to greatest in:
    - each row, left to right.
    - each column, top to bottom.

    0.063   1.00   0.73
    1.63   1.78   1.31
    0.037   1.97   0.31

1.7 Comparing and Ordering Decimals

1.8 Rounding Decimals

Objective: to round decimals to a given place value

Sometimes you may need to round a decimal to get an approximate value that is easy to use. Rounding decimals is just like rounding whole numbers.

Examples

A. In 2006, one British pound was worth 1.81812 U.S. dollars. Round the number of dollars to the nearest cent.

▶ 1.81812 is between 1.81 and 1.82.   1.8_1812   Underline the hundredths place. Look at the digit one place to its right.

If the digit is 5 or greater, round up. If the digit is 4 or less, the underlined digit stays the same.

To the nearest cent, 1.81812 rounds up to $1.82.

B. One kilometer equals 0.6213712 mi. Round this number to the nearest tenth.

▶ 0.6213712 is between 0.6 and 0.7.   0.6_213712   After you round a decimal number, drop the digits to the right of the place you are rounding to.

To the nearest tenth, 0.6213712 rounds to 0.6.

A kilometer is about six tenths of a mile.

Round each number to the underlined place-value position.

1. 3.6_21
2. 6.80_5
3. 9.6_128
4. 1.7_04
5. 1.7_04
6. 0._897
7. 3.76_0
8. 5.890_7

Exercises

Round to the nearest whole number.

1. 1.7
2. 3.27
3. 6.459
4. 7.3
5. 0.6
6. 4.803
7. 2.075
8. 9.94

Round to the nearest tenth.

9. 35.79
10. 35.35
11. 76.92
12. 99.345
13. $7.63
14. $99.64
15. 549.7
16. 900.09

Round to the nearest hundredth.

17. 8.0632
18. 3.9286
19. 0.0685
20. 6.8559
21. 7.908
★ 22. 8.9996

Problem Solving

23. Alice's fastest speed for running the 100-meter dash is 14.65 seconds. Round this to the nearest tenth of a second.

★ 24. What is the greatest number that rounds to 6.9? the least number?

Constructed Response

25. While shopping, Thomas selected a notebook for $1.95, a ruler for $1.09, a package of pens for $1.89, and an eraser for $0.79. Which two items cost about $1.00 each when rounded to the nearest dollar? Explain your reasoning.

Test Prep

26. How do you say 3.045?
    a. three and forty-five
    b. three and forty-five hundredths
    c. three and four thousandths
    d. three and forty-five thousandths

27. Which number is *not* greater than 8.076?
    a. 8.08
    b. 8.07
    c. 8.8
    d. 8.7

1.9 Metric Units of Measurement

Objective: to estimate and measure using metric measurements of length, capacity, and mass

The **metric system** is a base ten system. Most countries around the world use the metric system. The metric system is also used in science. The metric system missed being nationalized in the United States by one vote in the Continental Congress.

Metric units use a prefix and a base unit.

milli means thousandth
centi means hundredth
kilo means thousand

A cent is $\frac{1}{100}$ of a dollar.

A millimeter is about the thickness of a dime.

A centimeter is about the width of a fingernail.

A decimeter is about the length of a new crayon.

Metric Length

1 millimeter (mm) = 0.001 m
1 centimeter (cm) = 0.01 m
1 decimeter (dm) = 0.1 m
1 meter (m) = 1.0 m
1 kilometer (km) = 1,000 m

A doorknob is about 1 meter off the floor.

A kilometer is about five city blocks.

You can relate place values and metric units.

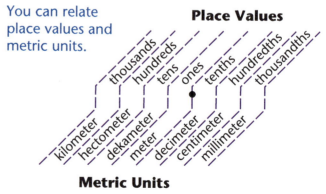

Place Values

Metric Units

20   1.9 Metric Units of Measurement

A grain of salt has a mass of about a milligram.

A small paperclip has a mass of about a gram.

A liter equals about 4 cups of liquid.

Metric Mass

1 milligram (mg) = 0.001 gram (g)
1 kilogram (kg) = 1,000 g
1 metric ton (t) = 1,000 kg

Metric Capacity

1 milliliter (mL) = 0.001 liter (L)

Your math book has a mass of about a kilogram.

A million dollars weighs a metric ton.

A milliliter equals about ten drops from an eyedropper.

TRY These

Choose the more reasonable estimate for each object.

1. the distance between two cities   a. 150 m   **b. 150 km**
2. the mass of a grape   **a. 5 g**   b. 5 kg
3. the capacity of a fishtank   a. 37 mL   **b. 37 L**
4. the length of a pencil   **a. 185 mm**   b. 185 cm
5. the mass of a truck   a. 3 kg   **b. 3 t**
6. the capacity of a medicine bottle   **a. 150 mL**   b. 150 L
7. the height of chalkboard   a. 3 dm   **b. 3 m**
8. the mass of a bag of apples   a. 1 g   **b. 1 kg**

Exercises

Complete using the correct unit of measurement.

1. 1 cm = 10 __mm__
2. $\frac{1}{1,000}$ kg = 1 _____
3. 1 m = 100 _____
4. 1,000 mL = 1 _____
5. 1,000 mg = 1 _____
6. $\frac{1}{10}$ m = 1 _____

Fill in each ____ with the missing metric unit for each object.

7. length of a pencil 14.8 ____
8. mass of a nickel 5 ____
9. length of your math book 26 ____
10. amount of soda in a bottle 2 ____
11. distance of a running race 8 ____
12. mass of a large truck 20 ____

Fill in each ____ with the correct metric measurement units.

13. Ben is 1.6 ____ tall and has a mass of about 39 ____. Fido, his dog, is 83 ____ long and weighs about 12 ____. Fido drinks an average of 3 ____ of water each day.

Problem Solving

14. Find the length of $\overline{XY}$ to the nearest centimeter and to the nearest millimeter.

Constructed Response

15. Liam Hanson will run the 100-m dash. Is this more or less than 1 km? Explain your reasoning.

Problem Solving

Checkmate

The small covering tile is exactly the same size as two squares on the checkerboard.

Cover the checkerboard using tiles so that every square is covered, but the tiles do not overlap. What is the greatest number of tiles that will fit on the board?

Using seven tiles, can you cover enough so that it is impossible to put another tile on the board?

Can you do it with six tiles? five tiles?

Extension

1. This checkerboard has four squares on each side.
   It is a 4 × 4 checkerboard.
   Make an 8 × 8 checkerboard.
   Make 2 × 2 tiles.

2. What is the greatest number of 2 × 2 tiles that will fit on the 8 × 8 board you made?

3. Using ten 2 × 2 tiles, can you cover enough squares so that it is impossible to put another tile on the 8 × 8 board? Can you do it with fewer tiles?

Problem Solving 23

Chapter 1 Review

LANGUAGE and CONCEPTS

Complete by using the words from the box.

1. The metric system is a _____ system.
2. The problem-solving plan has _____ steps.
3. Meters are a measurement of _____.
4. Grams are a measurement of _____.
5. Milliliters are a measurement of _____.
6. In the number 46,860,885, the 4 has a value of _____ million.

| capacity |
| forty |
| base ten |
| length |
| four |
| weight |
| mass |

SKILLS and PROBLEM SOLVING

Copy and complete. (Section 1.1)

7. 60 = ■ tens
8. 500 = ■ tens
9. 380 = ■ tens
10. $620 = ■ ten-dollar bills
11. 8,207 = ■ hundreds ■ ones

Write in standard form, expanded form, and words. (Section 1.2)

12. 222,222
13. 600,100,000
14. 83,000,000
15. eighty thousand, two hundred
16. seven billion, three thousand
17. 40,000 + 3,000 + 70 + 5
18. 200,000 + 90,000 + 400 + 80 + 1

Write the value of the underlined digit. (Section 1.2)

19. <u>2</u>4,371
20. 123,<u>4</u>56
21. 7,3<u>8</u>7,231
22. 5<u>7</u>,802,571
23. 62,5<u>0</u>2,317
24. <u>5</u>02,293,625

Compare using <, >, or =. (Section 1.3)

25. 75 ● 57
26. 809 ● 908
27. 3,912 ● 3,912
28. 500 ● 499
29. 2,371 ● 2,317
30. 4,362 ● 4,662

Order from least to greatest. (Section 1.3)

31. 140   410   401   104

32. 3,915   3,951   3,159

33. 58,713   75,238   64,919

34. 30,500   30,005   30,050

Choose the correct answer for rounding to the nearest thousand. (Section 1.4)

35. 6,128   **a.** 6,000   **b.** 7,000

36. 47,512   **a.** 47,000   **b.** 48,000

37. 9,542   **a.** 9,000   **b.** 10,000

38. 873,257   **a.** 873,000   **b.** 874,000

Round to the nearest ten thousand. (Section 1.4)

39. 73,199   **40.** 28,123   **41.** 61,999   **42.** 76,104   **43.** 308,000

Write the value of each underlined digit. (Section 1.6)

44. 5.8<u>7</u>   **45.** 3.98<u>7</u>   **46.** 19.0<u>5</u>1   **47.** 23.87<u>6</u>

Write each as a decimal. (Section 1.6)

48. six and four tenths

49. seven and fifty thousandths

Compare using <, >, or =. (Section 1.7)

50. 9.8 ● 9.80   **51.** 9.809 ● 9.810   **52.** 13.061 ● 13.060   **53.** 8.90 ● 8.09

Order from least to greatest. (Section 1.7)

54. 8.06   6.8   8.6   6.08

55. 5.29   5.9   5.02   5.09

Round each decimal to the underlined place-value position. (Section 1.8)

56. 7.8<u>9</u>5   **57.** 6.0<u>3</u>1   **58.** 1<u>0</u>.876   **59.** 1.9<u>3</u>2

Choose the more reasonable estimate for each object. (Section 1.9)

60. a glass of milk
  a. $\frac{1}{2}$ L   **b.** 2 mL

61. distance across a town
  a. 20 dm   **b.** 20 km

Solve. (Section 1.5)

62. Marcus lives 4.024 kilometers from school. Melanie lives closer to the school than Marcus does. Which could be the distance Melanie lives from the school, 4.025 km or 4.023 km? Explain your answer.

Chapter 1 Test

Write the value of the underlined digit.

1. 56,7<u>8</u>4
2. 4<u>1</u>3,725
3. <u>6</u>,730,020
4. 25,9<u>3</u>6,800

Write in standard form.

5. 18 tens
6. 30 hundreds
7. 523 ten-dollar bills
8. 7 thousands 56 ones   7,056
9. 802 hundreds 20 ones
10. eighty-nine hundredths
11. three and forty-two thousandths

Compare using <, >, or =.

12. 348,421 ● 348,412
13. 4,298,109 ● 4,289,901
14. 6.90 ● 6.9
15. 8.32 ● 8.321

Order from greatest to least.

16. 18.208   18.820   18.280
17. 29.435   92.345   29.534
18. 30,033   30,303   33,030   33,300
19. 51,115   51,511   51,151   51,115

Round to the underlined place-value position.

20. <u>4</u>.89
21. 6.0<u>5</u>9
22. 5.9<u>2</u>1
23. 3.7<u>3</u>5
24. 5<u>6</u>,919
25. 48,<u>1</u>09
26. <u>2</u>9,314
27. 71,6<u>3</u>9

Solve. 7000   30,000

28. Dwight D. Eisenhower was born in 1890. His brother Arthur was born in 1886, and his brother Edgar in 1889. Who was the oldest?

29. Mt. Blanc in the Alps is 15,771 feet high. To the nearest 1,000 feet, how tall is it?

30. Sarah ran a mile in 9.75 minutes. Sophie ran faster than Sarah. Which could be Sophie's time, 9.76 or 9.7 minutes? Explain your thinking.

9.76 because its bigger

Change of Pace

International Standard Notation of Time

Have you ever seen a clock like this?

Around the globe, most industrialized countries tell time using a 24-hour clock. In the United States and Canada, this is called military time. In Great Britain, it is referred to as continental time.

Using a 24-hour clock, time is told in hours and minutes (hh:mm). The day starts at midnight (00:00) and ends at 23:59. In this time, midnight is 00, 1 A.M. is 01, 1 P.M. is 13, and so on. Since there is a two-digit number for each of the 24 hours, there is no need for A.M. or P.M.

04:52 means 4:52 A.M. because 4 hours and 52 minutes have passed since midnight.

13:30 means it is 1:30 P.M. because 13 hours and 30 minutes have passed since midnight.

Try telling time using a 24-hour clock. Write the time as it would be shown on a standard clock. Include A.M. or P.M.

1. 15:00
2. 14:30
3. 23:25
4. 20:00
5. 16:45
6. 5:14
7. 23:00
8. 22:15
9. 17:40

Write each time using a 24-hour clock. Remember that you do not need to tell whether the time is A.M. or P.M.

10. 1:00 A.M.
11. 7:35 P.M.
12. 10:05 P.M.
13. 1:00 P.M.
14. 3:50 A.M.
15. noon

Cumulative Test

1. Which number is the standard form for six hundred thirty and three tenths?
   a. 630.3
   b. 6,030.30
   c. 60,030.310
   d. none of the above

2. Which digit is in the thousands place in the number 127,638?
   a. 1
   b. 6
   c. 7
   d. none of the above

3. What is 4.378 rounded to the nearest tenth?
   a. 4.3
   b. 4.38
   c. 4.4
   d. none of the above

4. Which replacement for ● makes this true?

   6.43 ● 6.34
   a. <
   b. >
   c. =
   d. none of the above

5. Which number is 1,000 greater than 235,891?
   a. 335,891
   b. 245,891
   c. 236,891
   d. none of the above

6. Which is the greatest number in the set of numbers 2.031, 2.301, 2.103, and 2.310?
   a. 2.310
   b. 2.103
   c. 2.301
   d. 2.031

7. What is the value of the 4 in 246,318?
   a. 400,000
   b. 40,000
   c. 4,000
   d. 400

8. Frank is 7 years older than his sister. She is 29 years younger than their father. Their father is 38. How old is Frank?
   a. 9
   b. 14
   c. 16
   d. 29

9. Dwight D. Eisenhower was the thirty-fourth president of the United States. Bill Clinton was the forty-second president. How many presidents served between them?
   a. 5
   b. 6
   c. 7
   d. 8

10. Choose the most reasonable estimate for the length of a large garden.
    a. 30 dm
    b. 30 m
    c. 30 cm
    d. 30 km

CHAPTER 2
Addition and Subtraction

Ian McAlister
Tolland, MA

2.1 Properties of Addition

Objective: to identify and use the properties of addition

Certain ideas are always true about how numbers work. These ideas are called **properties**. You can depend on properties always being true.

Commutative Property of Addition (Order)
Changing the order of the addends does not change the sum.

Examples

A. 25 + 167 + 75 = ?

You can add the numbers in order from left to right.

Or you can use the Commutative Property to change the order of the addends so you can use mental math. The sum is 267 either way.

▶ *Commutative sounds like commute, which means to go back and forth.*

THINK 25 + 75 equals 100.
25 + 75 + 167 = 267

> **Associative Property of Addition** (Grouping)
>
> Changing the grouping of three or more addends does not change the sum.

B. (63 + 25) + 25 = ?   113

You can add the numbers in the parentheses first.

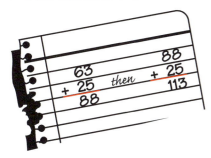

Or you can use the Associative Property of Addition to change the grouping so you can use mental math. The sum is 113 either way.

THINK 25 + 25 = 50
63 + (25 + 25) = 113

> **Identity Property of Addition** (Zero)
>
> Adding zero to any number does not change the number.

Identity numbers combine with other numbers, in any order, without changing the original number.

The identity number for addition is zero because when you add zero to any number, the sum is the number.

C. 43 + 0 = 43   **D.** 0 + 3.46 = 3.46

2.1 Properties of Addition

Addition Property of Equality

You can add to one side of the equal sign, but to keep the two sides equal, you must add the same amount to the other side.

Think of the equal sign like a scale. Both sides of the equation must be the same for the scale to stay in balance and for the equation to be true.

The amount in one pan balances the amount in the other pan. $3 = 3$	If you add an amount to one pan, the scale becomes unbalanced. $3 + 1 \neq 3$	If you add the same amount to the other pan, the scale becomes balanced again. $3 + 1 = 3 + 1$

TRY These

Match the letter of each property with the example shown.

1. $4 + 3 = 4 + 3$
2. $93 + 38 + 7 = 93 + 7 + 38$
3. $0 + 1.25 = 1.25$
4. $(27 + 15) + 15 = 27 + (15 + 15)$

a. Identity Property of Addition
b. Commutative Property of Addition
c. Associative Property of Addition
d. Addition Property of Equality

32 2.1 Properties of Addition

Fill in each blank and identify the property being used.

1. 67 + ____ = 67
2. 5 + 8 = 5 + ____
3. (17 + 98) + 2 = 17 + (_2_ + _98_)
4. 12 + 75 + 13 = 12 + ____ + ____
5. 6.85 + 12 = ____ + 12
6. 0.25 + 45 + 0.75 = ____ + ____ + ____
7. _3.5_ + 0 = 3.5
8. 48 + (5 + 12) = (48 + ____) + 5

Choose from the following numbers to write an example for each of the properties.

| 0 5 25 55 75 |

9. Commutative Property of Addition
10. Associative Property of Addition
11. Identity Property of Addition
12. Addition Property of Equality

(25 + 55) + 5 = 85

Problem Solving

Constructed Response

13. Jasper is trying to add up his most recent bowling scores of 87, 94, and 73. Explain how he could use the Commutative Property of Addition to help him find his sum.

2.1 Properties of Addition

2.2 Mental Math: Computing Sums and Differences

Objective: to use mental math to find sums and differences

Rhonda has read 34 pages of a book about Abraham Lincoln. She has 52 pages left to read. How many pages are in the book?

One way you can add or subtract mentally is to start at the greatest place-value position.

Add. 34 + 52   **THINK** 3 tens + 5 tens = 8 tens } 8 tens 6 ones = 86
    34 + 52 = 86   4 ones + 2 ones = 6 ones

There are 86 pages in the book.

Subtract. 86 − 34   **THINK** 8 tens − 3 tens = 5 tens } 5 tens 2 ones = 52
    86 − 34 = 52   6 ones − 4 ones = 2 ones

More Examples

Find other ways that make it easier for you to add.

A. Add. 7 + 8 + 3 + 2

    7 + 8 + 3 + 2 = 20

Look for a ten.
THINK 7 + 3 = 10 and 8 + 2 = 10
10 + 10 = 20

B. Add. 23 + 59

    23 + 59 = 82

Use compensation.
THINK 59 is one less than 60.
23 + 60 = 83, so one less is 82.

C. Subtract. 83 − 39

    83 − 39 = 44

Use compensation.
THINK 40 is one more than 39.
83 − 40 = 43, so one more is 44.

D. Subtract. 92 − 46

    92 − 46 = 46

Add the same digit to both numbers.
THINK (92 + 4) − (46 + 4)
96 − 50 = 46

TRY These

Solve mentally. Explain how you found each sum or difference.

1. 106 − 45
2. 45 + 54
3. 87 − 23
4. 6,700 + 2,200
5. 460 − 250
6. 36 + 79
7. 810 + 790
8. 156 − 92
9. 1,800 + 3,200
10. $0.99 − $0.49
11. 1 + 8 + 9 + 7 + 2
12. 80 − 69

Exercises

Solve mentally. Write only the sum or difference.

1. 32 + 64
2. 620 + 350
3. 580 + 210
4. 6,400 + 2,500
5. 13,000 + 42,000
6. 6 + 4 + 5 + 1
7. 5,900 + 1,300
8. 168 − 77
9. 94 − 51
10. 750 − 450
11. 137 − 95
12. $117 − $33
★ 13. 39,000 + 39,000
★ 14. 8,600 − 6,900
★ 15. 1,570 − 630
★ 16. $1,350 − $640

PROBLEM Solving

Solve mentally. Use the bar graph.

17. If one goal of the book club is for each member to read 150 books, how close is Carol to reaching the goal?
18. How many books have Roy and Vicki read in all?
19. If Norma reads 50 more books will she reach the goal?
20. How many more books has Carol read than Norma?
21. How many books have Vicki and Norma read in all?

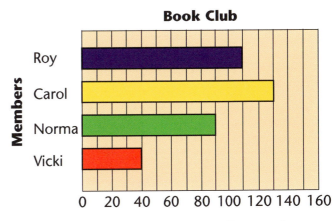

Book Club

2.2 Mental Math: Computing Sums and Differences

2.3 Mental Math: Estimating Sums and Differences

Objective: to estimate sums and differences

The bookstore had 1,010 books on sale. Of them, 877 were sold. About how many books were *not* sold?

To find the number of books not sold, subtract 877 from 1,010. Since you only need to know *about* how many books, you can **estimate**.

One way to estimate sums and differences is by rounding.

Round each number to the greatest shared place-value position. Then, add or subtract.

Estimate. 1,010 − 877

```
   1,010  →      1,000      1,010 rounds to 1,000
 −   877  →   −    900        877 rounds to 900
                   100   ←  estimate
```

About 100 books were not sold.

Examples

A.
```
    4,092  →    4,000
    1,989  →    2,000
 +  3,246  →  + 3,000
                9,000
```

B.
```
   $8.72  →    872  →    870
 + 0.74   →  +  74  →  +  70
                         940  → $9.40
```

Estimate dollars and cents like whole numbers.

TRY These

Estimate using rounding.

1.
```
    327
    449
 +  806
```

2.
```
    782
    334
 +  850
```

3.
```
    853
 −   48
```

4.
```
   $6.85
   13.62
 +  0.99
```

Estimate using rounding.

1. 752
   831
   + 955

2. 82¢
   − 59¢
   23

3. 3,68
   − 1.94
   174

4. $6.50
   + 2.75

5. 428
   − 83

6. 6.15
   + 1.98

7. 4,792
   1,320
   + 3,298

8. 1,128
   − 895

9. 7,690
   − 957
   33

10. 31,241
    + 2,980
    28,261

11. Estimate the sum of $47.85 and $28.13.

★ 12. Estimate how much greater 13,254 is than 9,089.

PROBLEM Solving

13. *Apollo 8* flew 550,000 miles on its trip around the Moon. *Apollo 9* flew 3,700,000 miles on its trip. About how many miles were flown by the two spacecraft?

14. In one year 28,762 new books were published. The next year 30,387 books were published. To the nearest thousand, how many books were published during the 2 years?

15. Ines brought 4 dozen donuts to the book club meeting. Later there were 15 donuts left. About how many were eaten?

16. Raquel needs 1 pair of shoes and 4 pairs of socks. A pair of socks costs $2.00 and a pair of shoes costs $15.99. Estimate how much money she will need.

MIXED Review

Solve.

17. Samuel wants to buy one book for $1.95 and another one for $4.99. Is $9.00 enough to pay for the books if there is a sales tax of $0.42?

18. How many quarters could you trade for 50 pennies? for 50 nickels? for 50 dimes?

Order from least to greatest.

19. 4,242   4,422   4,224

20. 50,050   50,505   50,005

21. 81,265   82,621   81,562

22. 35,004   35,084   35,048

2.3 Mental Math: Estimating Sums and Differences

2.4 Subtracting Greater Numbers

Objective: to subtract greater numbers

Mr. Cruz reported that $10,628 has been raised to buy a used van for the library. If the van costs $12,575, how much more money is needed?

$12,575 − $10,628 = ?   Estimate.
13,000 − 11,000 = 2,000

Subtract the same way in which you subtracted 2- and 3-digit numbers.

- Line up the digits by place value.
- Rename when necessary.
- Subtract.

Step 1	Step 2
Subtract ones and tens.	Subtract hundreds, thousands, and ten thousands.
 6 15 $\$12,5\cancel{7}\cancel{5}$ $-10,628$ $\overline{47}$	 1 15 6 15 $\$1\cancel{2},\cancel{5}\cancel{7}\cancel{5}$ $-10,628$ $\overline{\$1,947}$

$1,947 is the amount of money needed.

Based on the estimate of $2,000, the answer is reasonable.

Examples

A.
 11
 6 $\cancel{7}$ 13
$3\,2,\cancel{7}\cancel{2}\cancel{3}$
$-1\,2,6\,3\,4$
$2\,0,0\,8\,9$

B.
 4 9 9 10
$2\cancel{5},\cancel{0}\cancel{0}\cancel{0}$
$-1\,3,7\,2\,1$
$1\,1,2\,7\,9$

Regroup 500 tens as 499 tens and 10 ones.

TRY These

Subtract.

1. 2,631
    − 523

2. 6,371
    − 954

3. 3,929
    − 2,809

4. 8,791
    − 2,890

5. 7,628
    − 5,437

6. 9,522
    − 3,411

7. 38,000
    − 12,457

8. 50,004
    − 26,318

Exercises

Subtract.

1. 4,702
    − 1,526

2. 46,827
    − 5,651

3. 38,382
    − 19,266

4. 36,405
    − 14,294

5. 36,075
    − 21,524

6. 32,452
    − 15,735

7. 37,982
    − 19,395

8. 1,637 − 829

9. 51,520 − 35,630

★ 10. 225,000 − 117,565

PROBLEM Solving

Solve. Use the chart.

11. What is the difference in the 1980 populations of Green City and Centerville?

12. What is the difference between the 1990 populations of Ourtown and Newland?

13. By how much did the population of Green City increase from 1980 to 2000?

Population			
	1980	1990	2000
Ourtown	11,603	13,090	16,378
Centerville	13,609	17,004	21,075
Green City	6,601	6,323	7,982
Newland	8,040	10,690	15,100

2.4 Subtracting Greater Numbers

Problem Solving

Happy Holidays

During the holiday season members of the Diaz family exchange greetings by telephone. Each person will either call a relative or be called by a relative.

It will take one telephone call for Lucas to talk with his grandmother.

It will take three telephone calls for three relatives to talk with each other.

For four relatives to talk with each other, it will take six telephone calls.

How many telephone calls are needed for nine relatives to talk with each of the others?

Extension

1. How many telephone calls will be needed for twelve relatives to talk with each of the others?

2. If 105 telephone calls are made, how many relatives can talk with each of the others?

Cumulative Review

Write the value of the underlined digit.

1. 12,4<u>5</u>9,235   2. 4<u>7</u>8,309,043   3. 7,491,<u>2</u>86   4. 7<u>6</u>,420,118

   400,000

Write in words.

5. 245,300   6. 18.065   7. 3,465,000

Copy and complete.

8. 3,200 = ■ hundreds   9. 4,950 = ■ tens   10. 360,000 = ■ thousands

Compare using <, >, or =.

11. 6,892 ● 6,792   12. 29,600 ● 28,800   13. 3,259,850 ● 3,269,508

Round to the underlined place-value position.

14. 6,<u>5</u>21   15. <u>3</u>,068   16. <u>5</u>.55   17. 82,<u>9</u>63,300   18. <u>6</u>,590,281

Estimate.

19. 549 + 375
20. $7.83 + $2.37
21. 1,841 + 675
22. 842 + 93 + 406

Add or subtract.

23. 79,452
    + 8,639

24. 29,134
    + 86,455

25. 13,288
    + 5,988

26. 561 + 290   27. 327 + 4,694   28. 7,200 + 3,865

29. ²309
    − 285
    024

30. 610
    − 97

31. 719
    − 586

Solve.

32. Jamal had 2,340 mL of a chemical for science. He accidentally spilled some. When he remeasured he only had 1,753 mL left. How much did he spill?

33. Lucy ran a 10-km race. The next day she ran half that far at practice. How far did she run during the 2 days?

Cumulative Review 41

2.5 Problem-Solving Strategy: Guess and Check

Objective: to solve problems using the guess and check strategy

You can solve some problems by making a guess and then checking it.

Students in Miss Tuscany's class ordered school shorts. They ordered 28 pairs of shorts and spent $300. How many denim shorts did they order?

School Shorts
Denim $15
Flannel $9

1. READ You need to find the number of denim shorts sold. You know the total amount spent, the total number of shorts bought, and the cost for each kind of shorts.

2. PLAN
- First make a careful guess.
- Check to see if your guess is correct.
- If your guess is not correct, use the result to improve your next guess.
- Repeat this plan until you find the answer.

3. SOLVE Make a table to record your guesses.

Denim	Cost	Flannel	Cost	Total
10	$150	18	$162	$312 (too much)
5	$75	23	$207	$282 (too little)
8	$120	20	$180	$300 (correct)

They bought 8 denim shorts and 20 flannel shorts.

4. CHECK When you solved the problem, you showed that the students bought 8 denim shorts and 20 flannel shorts, or 28 shorts. This would cost $300. The answer fits the problem.

TRY These

1. The bus driver's age is an even number between 30 and 50. The sum of the digits is 4. How old is the bus driver?

2. Clark can buy pencils in packages of 12 or 20. He buys 3 packages and gets 52 pencils. How many packages of each kind did he buy?

Solve

1. Robbie opened a book and added the page numbers on the facing pages. The sum was 1,225. What were the page numbers?

2. Caroline is 4 years old. Her aunt is 6 times as old. How old will Caroline be when she is half as old as her aunt?

3. The sum of two numbers is 30. Their product is 216. What are the numbers?

4. If Megan receives $0.85 in change in 7 coins, what are the coins?

5. Paige bought a book and stationery for $26. She spent $4 more on the book than on the stationery. How much did each item cost?

6. William spent $36 for a soccer ball and a basketball. The basketball cost $14 more than the soccer ball. How much did each item cost?

MID-CHAPTER Review

Identify each addition property used.

1. $19 + 0 = 19$
2. $43 + 78 + 7 = 43 + 7 + 78$

Solve mentally.

3. $83 + 25$
4. $96 - 22$
5. $560 + 330$
6. $1,430 - 720$

Solve using the guess and check strategy.

7. A tuna salad sandwich costs $0.70 and an egg salad sandwich costs $0.60. Eleanor bought 5 sandwiches for $3.30. How many of each kind did she buy?

2.5 Problem-Solving Strategy: Guess and Check

2.6 Adding Decimals

Objective: to add decimals

Dara Torres finished in seventh place in women's 100-meter freestyle swimming. It took Dara 1.32 seconds longer than the winning time of 54.93 seconds. To find Dara's time, you add.

An estimate is about 56 seconds.

To add decimals, line up the decimal points so that the place-value positions are lined up. Then add as with whole numbers.

Step 1	Step 2	Step 3
Add the hundredths.	Add the tenths.	Add the ones and tens.

54.9**3**
+ 1.3**2**
———
    5
	5**4**.**9**3
	+ 1.**3**2
	———
	   .**2**5
	
	Rename 12 tenths as 1 one and 2 tenths.
		5**4**.93
		+ 1.32
		———
		56.25

Dara's time in the race was 56.25 seconds.   Compare the answer to the estimate.

Examples

A. 8.4   →   8.400
   + 0.615 → + 0.615
                ———
                9.015

▶ Attach zeros so that each decimal has the same number of decimal places.

B. 6.72
   + 0.34
   ———
    7.06

C. 12.100
     3.000
   + 0.572
   ———
    15.672

TRY These

Add.

1.   0.9
   + 0.8

2.   0.08
   + 0.96

3.   3.52
   + 2.1

4. $34.56
   + 47.08

5.   5.3
     7.9
   + 8.23

Exercises

Add.

1. 1.8 + 2.3	2. $1.65 + 8.48	3. 19.3 + 5.8	4. 26.7 + 14.5	5. 3.41 + 2.29
6. 0.87 + 1.2	7. 4.0 + 0.65	8. 2.43 + 1.9	9. 0.45 + 2.066	10. 9.32 + 8.791
11. 0.9 0.6 + 1.1	12. 7.5 8.19 + 3.859	13. 5.3 7.9 + 8.23	14. 12.34 9.57 2.03 + 8.31	15. 0.7 17.41 2.63 + 1.765

16. 7 + 8.1

17. 3.8 + 4.1

18. 0.55 + 6.066

19. 0.605 + 0.314

20. 20 + 8.95

21. 121 + 9.49

22. 44 + 0.187 + 6

23. 7.50 + 3.95 + 6.42

★ **24.** 1.806 + 0.008 + 0.910 + 4.881

★ **25.** 4 + 3.18 + 0.3 + 0.815

PROBLEM Solving

26. Alex writes 39.2, 47.6, and 58.9 on a piece of paper. Then he finds the sum of any two numbers at a time.
- How many different sums does he get?
- What is the greatest sum?
- What is the least sum?

27. Anne lives 0.45 km from school. The school is 1.186 km from Sophie's house. If Anne walks from her house to school and then to Sophie's house, how far will she walk?

2.6 Adding Decimals

2.7 Subtracting Decimals

Objective: to subtract decimals

A gymnast received a total score of 19.1 for her two routines. One of her scores was 9.35. What was her other score?

To find the gymnast's other score, subtract 9.35 from 19.1.   An estimate is 19 − 9 or 10.

- First line up the decimal points.
- Then subtract as with whole numbers.

```
  19.10     Write 19.1 as 19.10 to help
−  9.35     line up the decimal points.
```

Step 1	Step 2	Step 3
Subtract the hundredths.	Subtract the tenths.	Subtract the ones.
$\overset{0\ 10}{\cancel{1}\cancel{0}}$ 19.$\cancel{1}\cancel{0}$ − 9.35 ——— 5	$\overset{8\ \cancel{0}\ 10}{19.\cancel{1}\cancel{0}}$ − 9.35 ——— .75	$\overset{18\ \ 10}{\cancel{1}\cancel{9}.\cancel{1}\cancel{0}}$ − 9.35 ——— 9.75

The gymnast's other score was 9.75.   Compare this to the estimate, 10.

More Examples

```
A.    1.5        B.    0.681        C.    0.9⁴3¹³         D.    0.9¹⁰¹⁰
    − 0.3            − 0.300            − 0.408                − 0.833
    ─────           ───────            ───────               ───────
      1.2              0.381              0.535                 0.077
```

TRY These

Subtract.

```
1.    2.4      2.    6.95      3.    0.61      4.    0.417      5.    9.4
    − 0.9          − 3.50          − 0.44          − 0.230          − 8.36
```

Exercises

Subtract.

1. 9.3 − 7.1
2. 8.4 − 2.5
3. 6.3 − 4.6
4. $8.52 − 3.41

5. 6.39 − 0.55
6. 0.462 − 0.134
7. 73.1 − 25.9
8. 40.3 − 8.6

9. 0.96 − 0.90
10. 0.40 − 0.17
11. 2.84 − 0.90
12. 43.20 − 4.17

13. 3.594 − 1.361
14. 20.00 − 8.95
15. 5.934 − 2.737
16. 9.000 − 0.814

17. 7.8 − 5.1
18. 12 − 9.49
19. 7.35 − 2.53
20. $9.31 − $2.55
21. 4.52 − 1
22. 48 − 0.965
★ 23. (0.173 − 0.08) + 1.857
★ 24. 17 − (11.4 + 0.62)

Find the pattern. Write the next three decimals.

25. 0.2, 0.4, 0.6, ■, ■, ■
26. 6.5, 6.2, 5.9, ■, ■, ■
27. 0.082, 0.081, 0.08, ■, ■, ■
28. 2.01, 2.04, 2.07, ■, ■, ■
★ 29. 1.1, 2.3, 3.5, ■, ■, ■
★ 30. 1.2, 1.2, 2.4, 3.6, 6, ■, ■, ■

Problem Solving

31. Jack had a box containing 0.453 kg of salt. He used 0.26 kg of salt for a science experiment. How much salt was left in the box?

★ 32. 22.8 + ▲ = ●
72.6 − ▲ = ●

▲ = _____   ● = _____

2.7 Subtracting Decimals

2.8 Problem-Solving Strategy: More Than One Solution

Objective: to solve problems in more than one way

Mr. Cruzados sells souvenirs at the Olympics. A customer gives him $2.00 to pay for a booklet that costs $1.60. What coins could Mr. Cruzados give as change?

1. READ The cost and amount paid are given. You need to list the coins given as change.

2. PLAN Start with $1.60. Then add the value of one coin at a time until the total is $2.00.

3. SOLVE This problem has more than one answer. Two typical solutions follow.

A.
$1.60
$1.70
$1.75
$2.00

Mr. Cruzados could give a dime, a nickel, and a quarter as change.

B.
$1.60
$1.70
$1.80
$1.90
$2.00

He also could give four dimes as change.

4. CHECK Since $2.00 − $1.60 = $0.40, Mr. Cruzados should give $0.40 as change. Both solution A and solution B are correct.

48   2.8 Problem-Solving Strategy: More Than One Solution

List the coins for each amount. Give *two* solutions for each problem.

1. $0.31
2. $0.48
3. 95¢
4. 83¢
5. $0.27

Give *two* solutions for each problem.

1. A $10 bill is given for a purchase of $9.45. What coins could be received as change?

2. You buy 2 items for $2.34 each. What bills and coins could you give the clerk?

3. How old am I? Last year my age was a multiple of 5. This year my age is a multiple of 3. I am less than 35.

Use any strategy.

4. Rosa has a board that is 5.75 feet long. She cuts off 2.25 feet. What is the new length of the board?

5. Dmitri Sautin of Russia won the men's platform diving with a score of 692.34. Four years earlier Sun Shuwei of China won with 677.31 points. How many more points did Dmitri Sautin score?

6. When Pablo Morales of the U.S. won the 100-meter butterfly his time was 53.32 seconds. Denis Pankratov of Russia won four years later with a time of 52.27. How much faster was Denis Pankratov's winning time?

7. Lars Riedel of Germany won the discus throw with a distance of 227.66 feet. Twenty-four years earlier Ludvik Danek won with a throw of 211.25 feet. How much longer was Lars Riedel's throw?

8. When Brad Bridgewater won the 200-meter backstroke for the U.S. his time was 118.54 seconds. Twelve years earlier Rick Carey's winning time was 120.23 seconds. How much faster did Brad Bridgewater swim?

9. When Dan O'Brien won the decathlon it was with 8,824 points. Bruce Jenner won ten years earlier with 8,617 points. How many more points did Dan O'Brien have?

2.8 Problem-Solving Strategy: More Than One Solution

2.9 Elapsed Time

Objective: to find elapsed time

You can use addition and subtraction to help you figure out elapsed time.

Anthony went for a canoe ride at 11:10 A.M. His ride was over at 12:25 P.M. How long did his ride last?

To find how much time has elapsed, you can subtract the two times.

```
   12 h 25 min
 − 11 h 10 min
    1 h 15 min
```

▶ You can write the times using hours and minutes.

His ride was 1 hour and 15 minutes long.

Suppose a canoe trip starts at 7:00 A.M. on June 19 and ends at 5:00 P.M. on June 21. How many days and hours will the canoe trip last?

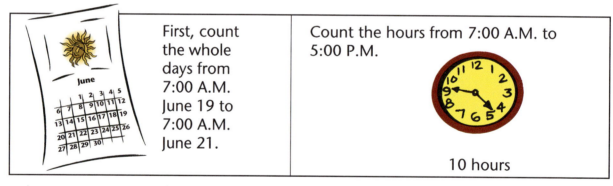

| First, count the whole days from 7:00 A.M. June 19 to 7:00 A.M. June 21. | Count the hours from 7:00 A.M. to 5:00 P.M.

10 hours |

So the trip will last 2 days and 10 hours.

TRY These

1. The movie starts at 1:25 P.M. and ends at 3:30 P.M. How long is the movie?
2. Kwan starts running at 3:35 P.M. If she runs for 45 minutes, when does she finish?
3. The Thomas family leaves for a trip at 4:00 P.M. on September 5. They return at 7:00 P.M. on September 9. How long was their trip?

Exercises

Find the time for each.

1. Start: 7:15 A.M.
   Elapsed Time: 3 h 18 min
   End: ?

2. Start: ?
   Elapsed Time: 1 h 35 min
   End: 12:00 P.M.

3. Start: December 7, 8:30 P.M.
   Elapsed Time: ?
   End: December 12, 12:00 P.M.

4. Start: March 15, 3:30 P.M.
   Elapsed Time: 9 d 17 h
   End: ?

	Activity	Starting Time	Ending Time	Elapsed Time
5.	School	8:05 A.M.	3:15 P.M.	
6.	Lacrosse practice	3:30 P.M.		1 h 55 min
7.	Dinner		6:50 P.M.	33 min
8.	Homework	7:10 P.M.	8:23 P.M.	

Problem Solving

9. There will be a 36-hour field trip to Antietam, Maryland, leaving from school at 7:45 A.M. on November 8. When will the students return to the school?

10. A train leaves New York at 12:55 P.M. It arrives in Westbury at 1:40 P.M. How long is the trip to Westbury?

★ 11. The fifth grade play will be on December 18. The cast wants to hang posters 2 weeks before the play. If it takes a week to make the posters, when should they start the posters?

★ 12. Heather wants to watch two movies back to back. The first one has a running time of 90 minutes and the second has a running time of 125 minutes. If she starts watching the movies at 6:30 P.M., when will she finish?

2.10 Problem-Solving Strategy: Exact Answer or Estimate

Objective: to determine whether an exact answer or estimate is appropriate

Carl Lewis's long jump measured 8.72 meters and Mike Powell's long jump measured 8.49 meters. Find the difference in the length of the jumps.

1. READ — You need to find the difference in the length of the jumps. You know that Lewis's jump was 8.72 meters and Powell's was 8.49 meters.

2. PLAN — You need to find an exact number. Subtract to find the difference.

3. SOLVE

```
  8.72
− 8.49
  0.23
```

The difference in the length of the jumps is 0.23 meters.

4. CHECK — Each jump was about 8 meters, so the answer must be less than 1.   0.23 < 1.00
The answer makes sense.

TRY These

Write *exact* if an exact calculation is needed or *estimate* if an estimate is enough.

1. Jan has saved $38.53. If he buys an Olympics calendar for $9.99 with $0.55 tax, about how much money will he still have?

2. To keep fit, Greg's goal is to swim 8 hours daily. He swam 6.5 hours before lunch, and 1.75 hours after lunch. Has he met his goal?

3. Of the 285 passengers on the jet, 138 were going to the Olympics. How many were not going there?

4. Tickets to the play were $22.00 each. About how much would 5 tickets to the play cost?

Solve

1. Greg Louganis got 730.80 points for men's springboard diving. Mark Bradshaw got 642.99 points. About how many more points did Greg get than Mark?

2. Allen Johnson ran the 110-meter hurdles in 12.95 seconds. One hundred years earlier it took Thomas Curtis 4.65 seconds longer. What was Curtis's time?

3. Janet Evans swam the women's 800-meter freestyle in 8 minutes, 20.20 seconds. That was 1.89 seconds less than the time it took the woman who won the silver. How long did it take the winner of the silver medal?

4. Gail Devers won two gold medals for the 100-meter run. Her first time was 10.82 seconds. Four years later her time was 10.94 seconds. How much faster did she run the first time?

5. The old shot put record of 36 feet 9.75 inches was broken. The new record is 73 feet 8.75 inches. About how many times greater is the new record?

6. Carey has saved $85.35 to buy sports gear. If he spends $42.72 for shoes, will he have enough money to buy sweat pants for $43.00?

7. A crowd of 5,623 people came to watch the basketball game in the afternoon. At night 16,365 people came to watch another game. About how many more people attended at night?

8. Some of the swimmers practiced freestyle for 3 hours and 45 minutes. Later they practiced the relays for 2 hours and 30 minutes. About how long did they practice in all?

MIND Builder

Patterns

A **palindrome** reads the same forward and backward.

eye   5005   73937   level

You can change any number into a palindrome.

Write any number.
Reverse the digits.
Add.
Continue until you find a palindrome.

```
   426
 + 624
  1050
 + 0501
  1551
```

Reverse the digits and add to find a palindrome.

1. 231
2. 85
3. 562
4. 645
5. 96
6. 192

Chapter 2 Review

LANGUAGE and CONCEPTS

Choose the letter of the property that matches each example.

1. 8.7 + 1.5 = 8.7 + 1.5
2. (29 + 85) + 15 = 29 + (85 + 15)
3. 0 + 4.005 = 4.005
4. 1.5 + 6 + 1.5 = 1.5 + 1.5 + 6

a. Identity Property of Addition
b. Associative Property of Addition
c. Addition Property of Equality
d. Commutative Property of Addition

SKILLS and PROBLEM SOLVING

Solve mentally. (Section 2.2)

5. 510 + 360
6. 16,000 + 32,000
7. 47 + 29
8. 3 + 6 + 4 + 2 + 7
9. 870 − 460
10. 8,900 − 5,600
11. $137 − $61
12. $1,420 − $720

Estimate using rounding. (Section 2.3)

13. 88 + 48
14. 47 − 31
15. 168 + 837
16. 694 − 423

17. 1.8 + 2.3
18. 16.7 − 8.4
19. 7.84 − 4.56
20. 5.12 + 6.31

21. 6,294 + 2,076
22. 8,710 − 3,462
23. $68.92 − 34.98
24. $58.03 + 47.61

Subtract. (Section 2.4)

25. 3,741 − 832
26. 51,375 − 9,278
27. 60,001 − 18,293
28. 72,067 − 13,072

54 Chapter 2 Review

Add or subtract. Sections (2.6–2.7)

29. 23.7
   − 14.9

30. 3.82
   + 4.06

31. 1.036
   + 0.777

32. 4.000
   − 1.621

33. 76.3 + 18.9

34. $1.74 − $0.27

35. 0.176 − 0.043

36. 0.38 + 1.9

37. 16 − 5.5

38. 0.261 − 0.09

Solve. (Sections 2.5–2.6, 2.8–2.10)

39. Danny buys a hot dog for $1.00, fries for 60¢, a soft drink for 95¢, and a piece of pie for $1.50. What is the total cost of his lunch?

40. Karen buys 3 kites for $1.79 each. What are *two* ways in which she could receive change from a ten-dollar bill?

41. Arlene is thinking of two numbers. They have a sum of 16 and a product of 48. What is the lesser number?

42. Letitia left for the beach at 4:15 P.M. on July 9. She returned home at 11:30 P.M. on July 17. How long was she gone?

43. Kenny worked a total of 12 hours last Saturday and Sunday at the pool. He worked two more hours on Saturday than on Sunday. How many hours did Kenny work each day?

Chapter 2 Test

Name the property used in each example.

1. $8.7 + 3 = 8.7 + 3$
2. $65 + 27 + 5 = 65 + 5 + 27$
3. $19 + 0 = 19$
4. $(36 + 90) + 10 = 36 + (90 + 10)$

Estimate using rounding.

5. $1.4 + 2.7$
6. $12.4 - 4.1$
7. $43{,}257 - 29{,}250$
8. $85{,}671 - 998$

Add or subtract.

9. $5{,}609 - 1{,}858$
10. $87{,}345 - 34{,}765$
11. $10{,}000 - 7{,}202$
12. $42{,}670 - 32{,}865$
13. $17{,}834 - 5{,}981$
14. $3.82 + 4.06$
15. $1.036 + 0.777$
16. $8 + 0.345$
17. $76.3 + 9.83$
18. $4.000 - 1.621$
19. $345.19 - 87.65$
20. $18 - 7.5$
21. $0.231 - 0.06$
22. $0.17 - 0.043$

Solve.

23. Alexis has 6 coins. The total value is $1.00. She has no pennies or half dollars. What coins does Alexis have and how many of each?

24. Shawn has basketball practice every afternoon starting at 3:40 P.M. until 5:15 P.M. How long is practice?

25. One serving of Pepe's pizza has 18 g of fat and 9.5 g of protein. How much more fat than protein is in one slice of Pepe's pizza?

26. What are *two* ways to receive change for a dollar and have at least one quarter and one dime?

Change of Pace

Roman Numerals

The Romans developed a set of numerals that can still be seen on some clocks and the cornerstones of some buildings. The Roman people of long ago did not use place value in their numeration as we do.

	1	5	10	50	100	500	1,000
Roman numerals	I	V	X	L	C	D	M

The symbols are written side by side. Sometimes the values of the symbols are added.

III $1 + 1 + 1$, or 3   **VI** $5 + 1$, or 6   **LXI** $50 + 10 + 1$, or 61

Subtract when a symbol is written before a symbol of greater value, as shown.

IV $5 - 1$, or 4   **IX** $10 - 1$, or 9   **XC** $100 - 10$, or 90

Write the number named by LXXXIV.

L   X   X   X   IV
↑   ↑   ↑   ↑   ↑
50 + 10 + 10 + 10 + 4

LXXXIV ⟶ 84

More Examples

Expanded form can help you write Roman numerals.

A. Write the Roman numeral for 48.

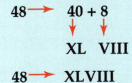

48 ⟶ XLVIII

B. Write the Roman numeral for 129.

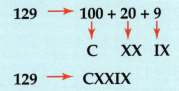

129 ⟶ CXXIX

Now Try These

Write the number named by each Roman numeral.

1. VII   2. XIV   3. XCI   4. CCXV   5. MMDXL

Write the Roman numeral.

6. 5   7. 20   8. 40   9. 61   10. 17

Cumulative Test

1. Which number is another way to write 45 tens?
   a. 45
   b. 4,510
   c. 45,000
   d. none of the above

2.
   ```
     286
     391
   + 475
   ```
   a. 945   b. 1,152
   c. 1,521   d. none of these

3. Estimate.
   ```
     10,385
   −  4,913
   ```
   a. 5,000   b. 6,000
   c. 7,000   d. none of these

4. Which set of numbers is in order from least to greatest?
   a. {6.5   6.4   6.3   6}
   b. {0.2   0.22   0.3   2}
   c. {1.8   1.5   2.0   0.8}
   d. { 3   3.1   3.03   3.33}

5. Ben has saved $18.00 toward a new ball glove. The price of the glove is $24.50. How much more must Ben save?
   a. $4.20
   b. $4.50
   c. $6.50
   d. none of the above

6. Kari ran the race in 14.56 seconds. How does she write the number in words?
   a. fourteen point fifty-six
   b. fourteen and fifty-six
   c. fourteen and fifty-six hundredths
   d. fourteen and fifty-six tenths

7. What is the *best* estimate for the amount of liquid in a tea cup?
   a. 250 mL
   b. 1 L
   c. 500 mL
   d. 25 mL

8. What is the *total* amount of money for all 5 days?

Day 1	$1.00
Day 2	$6.00
Day 3	$11.00
Day 4	$16.00
Day 5	$21.00

   a. $21.00
   b. $34.00
   c. $55.00
   d. none of the above

9. Joy has $10.00. She spends $2.85 for a sandwich and $5.82 for a book. Which fact is *not* needed to find how much money she spent in all?
   a. $10.00
   b. $2.85
   c. $5.82
   d. none of the above

10. What is 23,972 rounded to the nearest hundred?
    a. 23,072
    b. 23,900
    c. 24,000
    d. 24,900

CHAPTER 3

Multiplication

Sara Ludwig
Texas

3.1 Properties of Multiplication

Objective: to identify and use the properties of multiplication

The properties of multiplication are similar to the properties of addition. Both can help you compute mentally.

Commutative Property of Multiplication (Order)

Changing the order of the factors does not change the product.

$$5 \times 32 \times 2$$
$$5 \times 2 \times 32$$
$$10 \times 32 = 320$$

Associative Property of Multiplication (Grouping)

Changing the grouping of the factors does not change the product.

$$(16 \times 4) \times 25$$
$$16 \times (4 \times 25)$$
$$16 \times 100 = 1,600$$

Identity Property of Multiplication

If you multiply any number by one, the product is that number.

$$16 \times 1 = 16$$

Zero Property of Multiplication

If you multiply any number by zero, the product is zero.

$$16 \times 0 = 0$$

Multiplication Property of Equality

You can multiply both sides of an equation by the same number without putting it out of balance.

The amount in one pan balances the amount in the other pan.

$$2 = 2$$

If you multiply the amount in one pan by 2, the scale is unbalanced.

$$2 \times 2 \neq 2$$

If you multiply the amount in the other pan by 2, the scale balances again.

$$2 \times 2 = 2 \times 2$$

TRY These

Copy. Replace each ■ to make a true equation.
Write the name of the property that you used.

1. $3 \times 7 = \blacksquare \times 3$
2. $\blacksquare \times 7 = 0$
3. $1 \times \blacksquare = 8$
4. $(2 \times 2) \times 4 = 2 \times (\blacksquare \times 4)$
5. $4 \times \blacksquare = 4 \times 5$

Exercises

Without multiplying, match problems 1–17 with problems a–l that have the same product. The letters may be used more than once.

1. 26×4
2. $(63 \times 10) \times 10$
3. 48×1
4. $5 \times 2 \times 5$
5. $2 \times 5 \times 5$
6. $10 \times 54 \times 3$
7. 23×12
8. 93×27
9. 35×49
10. $700 \times 50 \times 10$
11. 32×16
12. $63 \times 0 \times 10$
13. 5×10
14. $2 \times 54 \times 5$
15. $1 \times 4 \times 26$
16. $2 \times 3 \times 8$
17. $7 \times 7 \times 7 \times 5$

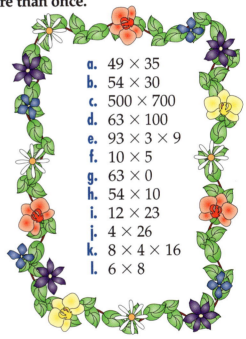

a. 49×35
b. 54×30
c. 500×700
d. 63×100
e. $93 \times 3 \times 9$
f. 10×5
g. 63×0
h. 54×10
i. 12×23
j. 4×26
k. $8 \times 4 \times 16$
l. 6×8

18. What number times 4,672 is 4,672?
19. What number times 189 is zero?

PROBLEM Solving

20. Mason orders 3 packages of bulbs from a flower catalog. Each package has 2 bulbs at a cost of $5 per bulb. How much will Mason pay for the order?

★21. Patrick earns $7 an hour. How much money does he earn for a 40-hour workweek?

3.2 Mental Math: Multiplying Multiples of 10, 100, and 1,000

Objective: to use facts and patterns to multiply mentally

Gilda's science team is trying to find the best kind of soil for cactus plants. How many plants do they have in all if each of the 6 students has 10 plants?

Since each student has the same number of plants, you can multiply to find the total.

THINK
$6 \times 1 = 6$
So, 6×1 ten $= 6$ tens
or $6 \times 10 = 60$

> Notice the number of zeros in the factors and the products. Is there a pattern?

There are 60 plants in all.

More Examples

A. 3×7 3×7 ones $= 21$ ones $= 21$
3×70 3×7 tens $= 21$ tens $= 210$
3×700 3×7 hundreds $= 21$ hundreds $= 2,100$
$3 \times 7,000$ 3×7 thousands $= 21$ thousands $= 21,000$

B. $5 \times 8 = 40$
$5 \times 80 = 400$
$5 \times 800 = 4,000$
$5 \times 8,000 = 40,000$

C. $9 \times 6 = 54$
$90 \times 60 = 5,400$
$90 \times 600 = 54,000$
$900 \times 600 = 540,000$

The pattern of zeros in example B is different from those in examples A and C because $5 \times 8 = 40$.

Use multiplication facts and patterns to find the products.

1. 2×5
2×50
2×500
$2 \times 5,000$
$2 \times 50,000$

2. 4×7
40×70
400×700

3. 8×8
80×80
800×800

Exercises

Multiply mentally. Write only the product.

1. 9 × 4,000
2. 5 × $20
3. 6 × 500
4. 80 × 4
5. 800 × 30
6. 30 × 900
7. 4,000 × 200
8. 70 × $70
9. 50 × 400
10. 300 × $60
11. 90 × 7,000
12. 200 × 200
13. (2 × 5) × 70
14. 8 × (60 × 5)
15. (8 × $5) × 400
★ 16. 60 × ■ = 4,800
★ 17. 300 × ■ = 21,000
★ 18. ■ × 90 = 6,300

Problem Solving

Solve. Use the graph for questions 19–21.

19. What was the total dollar amount of sales for the $6 planters?

20. What was the total dollar amount of sales for the $4 planters?

21. What was the total dollar amount of sales for the $20 planters?

Clay Planters Sale

Prices of Planters / Number of Planters Sold

Mixed Review

★ 22. Complete.

 and cost $8

 and cost $7

 and cost $12

 and cost ?

Round to the underlined place-value position.

23. 1,0̲99
24. 5̲,081
25. 17̲4,292
26. 83̲4
27. 4̲.365
28. 17.0̲9
29. 5̲.82
30. 215.39̲2

3.2 Mental Math: Multiplying Multiples of 10, 100, and 1,000

3.3 Mental Math: Estimation Using Rounding

Objective: to use rounding to multiply mentally

A boys' club ordered 325 trays of plants for a plant sale. About how many plants were ordered if each tray held 18 plants?

To find the total amount, multiply. You do not need an exact answer, so you may estimate.

One way to estimate products is by rounding.

- Round each factor to its greatest place-value position. Do *not* round 1-digit factors.
- Then multiply.

$$
\begin{array}{r} 325 \\ \times\ 18 \end{array} \rightarrow \begin{array}{r} 300 \\ \times\ 20 \\ \hline 6{,}000 \end{array}
$$

325 rounds to 300.
18 rounds to 20.

The club ordered *about* 6,000 plants.

More Examples

A.
$$
\begin{array}{r} 467 \\ \times\ 8 \end{array} \rightarrow \begin{array}{r} 500 \\ \times\ 8 \\ \hline 4{,}000 \end{array}
$$
467 rounds to 500.
1-digit factor

B.
$$
\begin{array}{r} 2{,}435 \\ \times\ 37 \end{array} \rightarrow \begin{array}{r} 2{,}000 \\ \times\ 40 \\ \hline 80{,}000 \end{array}
$$

C.
$$
\begin{array}{r} \$7.82 \\ \times\ 63 \end{array} \rightarrow \begin{array}{r} \$8.00 \\ \times\ 60 \\ \hline \$480.00 \end{array}
$$

D.
$$
\begin{array}{r} 3{,}229 \\ \times\ 438 \end{array} \rightarrow \begin{array}{r} 3{,}000 \\ \times\ 400 \\ \hline 1{,}200{,}000 \end{array}
$$

This estimate is low because the rounded numbers are both less than the original numbers.

TRY These

Estimate.

1. $\begin{array}{r} 36 \\ \times\ 9 \end{array}$ $\begin{array}{r} 40 \\ \times\ 9 \end{array}$

2. $\begin{array}{r} \$3.85 \\ \times\ 2 \end{array}$ $\begin{array}{r} \$4.00 \\ \times\ 2 \end{array}$

3. $\begin{array}{r} 328 \\ \times\ 36 \end{array}$ $\begin{array}{r} 300 \\ \times\ 40 \end{array}$

4. $\begin{array}{r} 5{,}897 \\ \times\ 71 \end{array}$ $\begin{array}{r} 6{,}000 \\ \times\ 70 \end{array}$

5. $\begin{array}{r} 708 \\ \times\ 190 \end{array}$ $\begin{array}{r} 700 \\ \times\ 200 \end{array}$

6. $\begin{array}{r} 4{,}212 \\ \times\ 359 \end{array}$ $\begin{array}{r} 4{,}000 \\ \times\ 400 \end{array}$

Exercises

Estimate.

1. 7,821 × 3
2. $62.79 × 8
3. 4,971 × 6
4. 9,721 × 7
5. 2,818 × 4
6. 16 × 44
7. 64 × 32
8. $57 × 76
9. 97 × 35
10. 81 × 23
11. 413 × 112
12. 776 × 317
13. 609 × 762
14. 557 × 842
15. $2.92 × 360

★ 16. 2,417 × 590
★ 17. 6,034 × $6.72
★ 18. 8,901 × 443

Estimate. Then choose the product that makes the most sense.

19. 189 × 4   a. 436   b. 510   c. 756
20. 27 × 9   a. 81   b. 243   c. 723
21. 34 × 29   a. 90   b. 986   c. 1,153
22. 56 × 25   a. 1,000   b. 1,217   c. 1,400

PROBLEM Solving

23. The members of the boys' club sold 2,615 plants. About how much did they collect if each plant sold for $6?

24. A nursery had 1,657 trees. They sold 798 trees. About how many trees were unsold?

25. There are 22 boys in Mr. Smith's sixth grade class. If he wants to give each boy 8 pencils, about how many pencils should Mr. Smith order?

3.3 Mental Math: Estimation Using Rounding

3.4 The Distributive Property

Objective: to use the Distributive Property to multiply

Steve is planting a garden in his backyard. He roped off an area that is 8 feet wide and 14 feet long. You can use simple multiplication facts to find the area of his garden.

First, use graph paper to draw a rectangle that is 8 units wide by 14 units long.

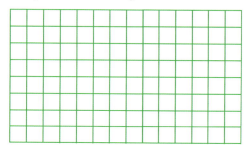

Each unit represents 1 foot.

The area equals 8 x 14.

You can divide the rectangle into two parts. Try to divide it so that there are groups of ten.

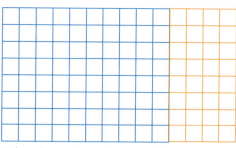

Why does it make sense to divide the rectangle to get groups of ten?

The area of the whole rectangle can now be found by adding the areas of the two smaller rectangles. When you do this, you are using the **Distributive Property**.

Area = (8 x 10) + (8 x 4)
Area = 80 + 32 (partial products)
Area = 112 sq feet

> To distribute means to spread. The Distributive Property lets you spread numbers out so they are easier to multiply.
>
> The Distributive Property says that you can multiply a sum by multiplying each addend separately. This gives you the **partial products**. Then add the products.

Another Example

Use the Distributive Property to find the product of 6 x 34.

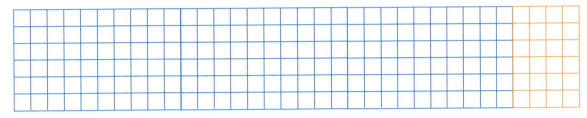

(6 x 30) + (6 x 4)

180 + 24 = 204

The product is 204.

List the partial products for each and find their sums. Then write a multiplication equation for each.

1.

2.

3.

4.

3.4 The Distributive Property

Exercises

On grid paper, draw and divide a rectangle to show each product. Use the Distributive Property to find the product.

1. 5 × 14
2. 9 × 15
3. 8 × 34
4. 3 × 28
5. 6 × 29
6. 4 × 37

Problem Solving

Use the Distributive Property to solve each problem.

7. Steve goes to the garden store to buy vegetable plants to put in his garden. Each tray has 6 plants. If he buys 14 trays, how many plants will he buy?

8. In the garden, there are 9 tomato plants. Each plant produced 14 tomatoes. How many tomatoes did Steve grow?

Constructed Response

9. Steve grew many zucchini. He made 3 loaves of zucchini bread for each of his 18 friends. How many loaves of zucchini bread did he make? Explain.

Mixed Review

Add or subtract.

10. 3.51
   + 2.83

11. 7.1
   − 3.5

12. 17.6
   + 1.9

13. 4.300
   − 1.851

Mind Builder

Logical Reasoning

Copy and complete.

1. 4,172
   ×   4
   ■,68■

2. ■,1■3
   ×   5
   10,815

3. 3,2■9
   ×   3
   9,■07

4. 16,■■8
   ×   7
   112,266

68 3.4 The Distributive Property

Problem Solving

Please Don't Squeeze the Oranges

A grocer wants to build a tower of oranges. The base of the tower is made of a square of 36 oranges. If each layer of the tower is a square, but smaller than the layer below it, how many oranges are needed to build the tower?

Extension

1. If the grocer used 385 oranges, how many oranges would be in the bottom layer?

2. If he used 650 oranges, how many layers would the tower have?

Problem Solving 69

3.5 Multiplying by Multiples of 10 and 100

Objective: to multiply whole numbers by multiples of 10 and 100

If you traveled in a car at a speed of 65 mph for 30 hours, how far would you travel?

To determine this, you would need to multiply 65 by 30. You can break down the multiple of 10.

65 × 30 = 65 × 3 × 10

```
  65
×  3
 ───
 195
```

- Then, multiply the product by 10.   195 × 10 = 1,950

- You can multiply by multiples of 10 in one step.

- Since 30 is a multiple of 10, the ones digit in the product must be 0.

> If one of the factors is a multiple of 10, why is the digit in the ones place always 0?

You can write a 0 in the ones place and then multiply 65 by 3.

```
   65
×  30
─────
1,950
```

Another Example
Multiply. 822 × 400

```
    822
×   400
───────
328,800
```
Since 400 is a multiple of 100, the digits in both the ones place and the tens place will be 0.

Multiply.

1. 87
  × 50

2. 879
  × 30

3. 438
  × 70

4. 689
  × 400

5. 604
  × 600

Exercises

Multiply.

1. 67 × 10
2. 45 × 30
3. 87 × 70
4. 38 × 50
5. 405 × 60

6. 662 × 80
7. 317 × 40
8. 123 × 90
9. 2,345 × 30
10. 3,405 × 70

11. 435 × 800
12. 814 × 400
13. 905 × 500
14. 1,234 × 200

Problem Solving

15. A family went on a road trip. If they drove for 36 hours at an average rate of 50 mph, how far did they travel?

16. A supermarket display is made by stacking cans of soup. If there are 30 cans stacked in a row, and there are 35 rows, how many cans are used to make the display?

Mid-Chapter Review

Copy. Replace each ■ to make a true equation. Write the name of the property that you used.

1. 452 × ■ = 452
2. 40 × 50 = ■ × 40
3. 7 × ■ × 65 = 0

Estimate.

4. 87 × 5
5. 392 × 8
6. 492 × 68
7. $3.95 × 48
8. 2,417 × 30

Multiply.

9. 712 × 50
10. $9.52 × 60
11. 2,548 × 200
12. 7,031 × 80
13. 3,000 × 50

3.5 Multiplying by Multiples of 10 and 100

3.6 Multiplying by 2-Digit Numbers

Objective: to multiply by 2-digit numbers

Mr. Becker's science class is collecting leaves from trees and plants. Each of the 24 students needs 12 different leaves. How many leaves will they have for the classroom display?

Estimate.
10 × 20 = 200

Step 1	Step 2	Step 3
Multiply by the ones.	Multiply by the tens.	Add the products.
24 × 12 ――― 48 2 × 24	24 × 12 ――― 48 240 10 × 24	24 × 12 ――― 48 + 240 ――― 288

THINK

12 × 24
(2 + 10) × 24
(2 × 24) + (10 × 24)
48 + 240 = 288

The students will collect 288 leaves for the classroom display.

Compared to the estimate, the answer is reasonable. Because both factors are rounded down, the estimate is low.

Examples

A.   35
  ×   42
  ―――
    70
+ 1,400
―――
1,470

B.   308
  ×   67
  ―――
  2 156
+ 18,480
―――
20,636

C. $5.28
×   53
―――
  15 84
+ 264 00
―――
$279.84

TRY These

Multiply.

1. 82 × 34
2. $32 × 41
3. 36 × 29
4. 902 × 79
5. $5.12 × 49
6. 132 × 37
7. 43 × 28
8. 213 × 42
9. $1.36 × 17

Exercises

Multiply.

1. 52 × 13
2. 64 × 12
3. $62 × 47
4. 93 × 15
5. 68 × 83
6. $84 × 65
7. 75 × 23
8. 48 × 34
9. 812 × 53
10. 605 × 61
11. 370 × 29
12. $2.28 × 38
13. 809 × 63
14. 646 × 39
15. 345 × 24

16. 18 × 19
17. 91 × 325
18. 62 × $3.40
★ 19. 1,151 × 74
★ 20. 92 × 2,091
★ 21. 6,154 × 17
★ 22. (25 × 6) × (8 × 8)
★ 23. (15 × 3) × (6 × 4)

Problem Solving

24. A gardener planted 21 palm trees in each row. How many trees did he plant if he planted 15 rows?

25. Julie worked 25 hours a week for 12 weeks this summer. How many hours did she work?

★ 26. An average fifth grader is 11 years old. How many months old is an average fifth grader?

Constructed Response

27. A supermarket display is made by stacking cans of soup. If there are 16 cans stacked in a row, and there are 35 rows, how many cans are used to make the display? Show all work and explain.

3.6 Multiplying by 2-Digit Numbers

3.7 Multiplying by 3-Digit Numbers

Objective: to multiply by 3-digit numbers

A nursery donated pine seedlings to 236 schools. Each school received 125 seedlings. How many seedlings were donated in all?

Multiply 236 and 125.

Estimate.
```
    200
 ×  100
 20,000
```

Step 1	Step 2	Step 3
Multiply by the ones.	Multiply by the tens.	Multiply by the hundreds. Then add the products.
$\begin{array}{r} \overset{1\ 3}{236} \\ \times\ 125 \\ \hline 1180 \end{array}$ 5×236	$\begin{array}{r} \overset{1}{\cancel{}\cancel{3}} \\ 236 \\ \times\ 125 \\ \hline 1180 \\ 4720 \end{array}$ 20×236	$\begin{array}{r} \overset{1}{\cancel{}\cancel{3}} \\ 236 \\ \times\ 125 \\ \hline 1180 \\ 4720 \\ +23600 \\ \hline 29,500 \end{array}$ 100×236

29,500 seedlings were donated.

Since both factors in the estimate were rounded down, the estimate is low. The answer is reasonable.

Examples

A.
```
      127
  ×   234
      508    (4 × 127)
    3 810    (30 × 127)
 + 25 400    (200 × 127)
   29,718
```

B.
```
      213
  ×   521
      213
    4 260
 + 106 500
   110,973
```

C.
```
      878
  ×   409
    7 902
 + 351 200
   359,102
```

D.
```
     $4.52
  ×   570
    316 40
 + 2 260 00
  $2,576.40
```

▲ Remember to show dollars and cents with two decimal places.

TRY These

Multiply.

1. 385 × 197
2. 592 × 388
3. 514 × 467
4. 781 × 908
5. $1.56 × 250

Exercises

Multiply.

1. 314 × 250
2. 681 × 142
3. 725 × 351
4. 162 × 925
5. 345 × 636
6. 835 × 649
7. 349 × 750
8. 871 × 536
9. 673 × 422
10. $2.01 × 904

11. 723 × 654
12. 999 × 888
13. $10.75 × 2.58

PROBLEM Solving

14. An orange grower shipped 240 boxes of oranges. Each box contained 145 oranges. How many oranges were sent?

15. An airplane travels at an average rate of 310 miles per hour. When it has reached 290 hours of flying time, how many miles has it flown?

16. Mars takes about 687 Earth days to complete an orbit around the Sun. How many Earth days would it take to complete 481 orbits?

17. The bakery sells pies for $8.99. If it sells 425 pies a month, how much money is collected from selling pies?

Constructed Response

18. The School Zone sells sets of encyclopedias for $2,000. Each set has 18 volumes. Each volume is sold individually for $115. Is it a better decision to purchase each volume individually, or to buy a set? Explain.

3.7 Multiplying by 3-Digit Numbers

3.8 Problem-Solving Strategy: Find a Pattern

Objective: to use patterns to solve problems

Celeste's mother, a freelance artist, has to make 42 sketches of animals for a science book. She decides to draw 2 sketches on Monday, 4 on Tuesday, and 6 on Wednesday. On which day will she complete her assignment?

▶ Assume that the pattern continues.

1. READ — You need to know the day on which she will complete her assignment. You know the total number of sketches she needs and the number she will draw on some of the days, but you need to find the number of sketches she will draw on the other days.

2. PLAN — Organize your data in a chart. Look for a pattern that will help you.

Days	M	T	W	Th	F	Sa
Sketches	2	4	6			
Total	2	6	12			

3. SOLVE — Complete the pattern.

Days	M	T	W	Th	F	Sa
Sketches	2	4	6	8	10	12
Total	2	6	12	20	30	42

Celeste's mother will complete her assignment on Saturday.

4. CHECK — 2 + 4 + 6 + 8 + 10 + 12 = 42
She will complete 42 drawings in 6 days, Monday through Saturday.

TRY These

Complete each pattern.

1. 15, 16, 14, 17, 13, 18, ■

2. 2, 4, 8, 16, 32, ■

Solve.

3. How many steps could be made with 28 blocks?

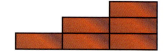

Solve. Assume each pattern continues.

1. Celeste's mother earned $100 for Monday's sketches, $300 for Tuesday's, and $500 for Wednesday's sketches. What were her total earnings for the 6 days?

2. At the first stop 25 people got on a train. At the next stop 40 people got on, and at the third stop 55 people got on. After how many stops were 275 people on the train?

3. Find a pattern in the design. How many squares will be in the twelfth design?

4. Crystal Clean Carwash gives one free wash every fifth wash. If you wash your car every week for a year, how many free washes will you get?

5. A square can be made with 4 toothpicks. It takes 7 toothpicks to make two squares side by side. Three squares in a row can be made with 10 toothpicks. How many toothpicks would you need to make four squares in a row?

★ 6. Ms. Tuscany's car had 60,000 miles on it and was valued at $12,500. When it reached 70,000 miles, it was valued at $11,750. At 80,000, the value dropped to $11,000. How much will the car be worth at 150,000 miles?

MIND Builder

Number Puzzles

$$\begin{array}{r} 6{,}532 \\ -\ 2{,}356 \\ \hline 4{,}176 \end{array}$$

1. Choose any four digits from 0 to 9.   5 3 2 6

2. Arrange them to make the greatest number possible.

3. Arrange them to make the least number possible.

$$\begin{array}{r} 7{,}641 \\ -\ 1{,}467 \\ \hline 6{,}174 \end{array}$$

4. Subtract.

5. Take the answer and repeat steps 2–5 until you get 6,174.

6. Do the puzzle using these digits.
    a. 3 5 4 6   b. 1 2 3 4   c. 5 1 7 3   d. 2 8 0 3

3.9 Multiplying Decimals by 10, 100, and 1,000

Objective: to multiply decimals by 10, 100, and 1,000

A runner ran in 10 races during the year. If each race was 6.4 miles long, how far did she race that year?

6,400 × 10 = 64,000
  640 × 10 = 6,400
    64 × 10 = 640
   6.4 × 10 = 64
 0.64 × 10 = 6.4

Remember that each place in the number system is 10 times greater than the place to its right.

When you multiply by 10, you are making a number 10 times greater. You can move the decimal point one place to the right.

Examples

A. 3.2 × 10 = 32

B. 0.05 × 10 = 0.5

C. 39.0 × 10 = 390

Think of whole numbers as having decimal points after the number. (39 = 39.0)

When you multiply by 100, you are making the number 100 times greater. You can do this by moving the decimal point two places to the right.

More Examples

D. 3.20 × 100 = 320

E. 0.05 × 100 = 5

F. 39.00 × 100 = 3,900

THINK How could you make a decimal 1,000 times greater?
4.543 × 1,000 = ?

That's right. 4,543

Move the decimal point three places to the right. One place for every 0 in 1,000.

Add a zero to hold a place if you need to.

78   3.9 Multiplying Decimals by 10, 100, and 1,000

Multiply by moving the decimal point. Remember that you can add zeros if needed.

1. 7.8 × 10
2. 14.56 × 10
3. 8.0 × 10
4. 7.05 × 100

Multiply by moving the decimal point. Remember that you can add zeros if needed.

1. 0.2 × 10
2. 13.08 × 10
3. 0.087 × 10
4. 8.5 × 10
5. 1.4 × 100
6. 0.987 × 100
7. 3.1 × 100
8. 0.076 × 100
9. 0.9 × 1,000
10. 3.145 × 1,000
11. 0.87 × 1,000
12. 145.9 × 1,000

Find the missing number.

13. 1.4 × _____ = 14
14. 0.87 × _____ = 87
15. 3.14 × _____ = 31.4
16. 0.1 × _____ = 1
17. 10 × _____ = 65.4
18. 10 × _____ = 0.54

Compare using <, >, or =.

19. 1.2 × 10 ● 0.12 × 100
20. 0.65 × 100 ● 6.3 × 10
21. 0.030 × 1,000 ● 0.3 × 10
22. 0.05 × 10 ● 0.005 × 1,000

PROBLEM Solving

23. An average jar of peanut butter is made from 5.48 × 100 peanuts. How many peanuts are used?

24. Joy bought 10 packages of chewing gum for $0.89 each. How much did she spend?

Constructed Response

25. If a penny weighs about 2.5 g, could you lift $100 worth of pennies? Explain your answer.

3.9 Multiplying Decimals by 10, 100, and 1,000

3.10 Multiplying Whole Numbers and Decimals

Objective: to find the product of a whole number and a decimal

The Roberts family is redecorating their basement. Two of the walls in the basement were covered with wallpaper that they are removing. Only four tenths of each wall still has wallpaper. What part of both walls is covered with wallpaper?

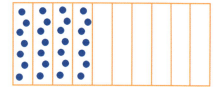 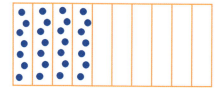

There is wallpaper remaining on 0.4 of the 2 walls.

$$0.4 \times 2$$

The word *of* indicates multiplication in math.

When you divide the walls into tenths, you see that there are 8 tenths, or 0.8, sections with wallpaper.

The product 0.8 is less than 2 because you are multiplying 2 by a factor that is less than 1.

More Examples

Multiply the problems just like whole numbers. Then use the decimal places in the factors to place the decimal in the product.

A. Multiply. 30×0.5

You are finding 5 tenths of 30.
5 tenths is $\frac{1}{2}$.
What is $\frac{1}{2}$ of 30?

$$
\begin{array}{r}
30 \\
\times\ 0.5 \\
\hline
15.0
\end{array}
$$
← 0 decimal places
← 1 decimal place
← 1 decimal place

B. Multiply. 1.25 × 8

$$\begin{array}{r} 1.25 \\ \times\ \ 8 \\ \hline 10.00 \end{array}$$ ← 2 decimal places
← 0 decimal places
← 2 decimal places

> Look at the number of decimal places in the product in both problems. This matches the number of decimal places in the factors.

TRY These

Draw models to show each of the following problems.

1. 2 × 0.9
2. 9 × 0.5
3. 16 × 0.1
4. 10 × 0.6

Exercises

Multiply. Use the number of decimal places in the factors to place the decimal point in the product.

1. 58 × 0.5
2. 24 × 0.25
3. 90 × 0.33
4. 1.75 × 40

Insert a decimal point to make each problem correct.

5. 42 × 0.035 = 1470
6. 4.73 × 188 = 88924
7. 8 × 8.188 = 65504
8. 0.42 × 136 = 5712
9. 100 × 0.125 = 12500
10. 2,354 × 1.34 = 315436

Multiply.

11. 14 × 2.6
12. 200 × 4.5
13. 0.35 × 8
14. 14 × 2.5

15. 22 × 1.1
16. 100 × 0.67
17. $0.05 × 5
18. 0.345 × 7

19. 1.365 × 2
20. 2.562 × 4
21. $35.62 × 8
22. 5.515 × 2

3.10 Multiplying Whole Numbers and Decimals

PROBLEM Solving

23. Kenny biked in a race that was 35 miles long. How many kilometers is this? (To convert miles to kilometers, multiply the miles by 1.61.)

24. A peregrine falcon can fly at a speed of 200 mph. How many km per hour is this?

Constructed Response

For problems 25–26, use the numbers shown. Put one number in each box.

★ **25.** Make the greatest possible product. Explain how you chose your numbers.

★ **26.** Make the least possible product. Explain how you chose your numbers.

4 5 6 7 8

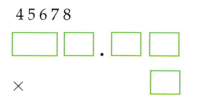

TEST Prep

27. Molly bought 5 shirts, each costing $8.99. There was no tax. What was the total cost of the shirts?

a. $449.50   b. $4.49   c. $45.00   d. $44.95

28. The pool pays its employees $7.25 an hour. Which expression would show the amount of money the employees earned in 8 hours?

a. $7.25 + 8

b. $7.25 × 8

c. $7.25 ÷ 8

d. $7.25 × 8 + 8

Cumulative Review

Write the value of the underlined digit.

1. 3,432,0<u>9</u>6
2. <u>7</u>,004,832
3. 6,93<u>5</u>,231
4. 9,<u>7</u>03,214
5. 8,149,<u>7</u>20
6. 5,0<u>3</u>0,629

Round to the underlined place-value position.

7. <u>6</u>,291
8. <u>8</u>,705
9. 3.<u>1</u>9
10. 4.5<u>5</u>

Order from least to greatest.

11. 3.05   5.03   3.15   5.3
12. 222   222,100   221,500   221,050

Copy and complete.

13. 700 = ■ tens
14. 4,200 = ■ hundreds
15. 65,000 = ■ thousands

Add or subtract.

16. 9.20 + 3.69
17. 8.00 + 0.56
18. 12.96 + 7.30
19. 10.0 − 2.7
20. 12.30 − 5.85

Multiply.

21. 51 × 7
22. 36 × 8
23. 268 × 5
24. $12.05 × 4
25. $7,543 × 6

26. 72 × 9
27. 861 × 2
28. $7.28 × 7

Solve.

29. Two painters were painting a house. Each painter was paid $16 an hour. The first painter worked 8 hours and the second worked 9 hours. How much total were the painters paid?

30. The Tour de France covers a distance of 3,607 km. Lance Armstrong won the race 7 times. How many kilometers total did he cover in his winning races?

3.11 Multiplying Decimals

Objective: to find the product of two decimal factors

Mrs. Bartell works at the newspaper office. She earns $6.20 an hour. If she works 7.5 hours a day, how much does she earn in one day?

Multiply to find how much she earns.

Use these steps when you multiply a decimal by a decimal.

Step 1	Step 2	Step 3
Estimate.	Multiply as with whole numbers.	Use the estimate to place the decimal point in the product.
$6.20 → $6 × 7.5 → × 8 ——— $48	$6.20 × 7.5 ——— 3100 4340 ——— 46500	$6.20 } 3 decimal × 7.5 } places ——— 3100 4340 ——— 46.500 3 decimal places

Mrs. Bartell earns $46.500 or $46.50 a day. Compared to the estimate, the product is reasonable.

Why is $46.500 equal to $46.50?

More Examples

A.
```
            Estimate.
   1.3  →       1
 × 3.5  →     × 4
 ————        ————
   6 5          4
   3 9
 ————
 4.5 5
```
Two decimal places make the product close to the estimate.

B.
```
              Estimate.
    7.2 0  →       7
  ×   9.5  →     × 10
  ——————         ————
    3 6 0 0        70
  6 4 8 0
  ————————
  6 8.4 0 0  or 68.4
```

84 3.11 Multiplying Decimals

TRY These

Estimate. Then copy the product. Use your estimate to help you place the decimal point in the product.

1.
```
    3.8
  × 2.6
   228
    76
   988
```

2.
```
    4.60
  ×  5.2
     920
    2300
   23920
```

3.
```
    1.19
  ×  1.7
     833
     119
    2023
```

4.
```
    0.9
  × 6.3
     27
     54
    567
```

Exercises

Multiply.

1. 6.1 × 4.2
2. 3.14 × 1.8
3. 1.3 × 5.1
4. 8.70 × 2.5
5. 1.8 × 1.8
6. 7.01 × 2.0
7. 6.5 × 4.9
8. $9.80 × 1.00

9. What is 1.3 times 9.8?
10. Find the product of 2.8 and $48.50.

PROBLEM Solving

11. If Mr. Blair drives 13.47 miles round trip to and from work each day, how far will he drive in 6.5 work days?

12. If a line of advertising in a newspaper costs $3.46, how much would you pay for 25.3 lines?

13. Marlene saved ten dollars worth of quarters. How many quarters did she save? Find *two* ways to solve this problem.

14. In science class each pair of students was given weights measuring 4.25 grams. How much weight was given to 6 pairs of students?

15. Judy rode her bicycle around the park track 3.5 times. How far did she ride?

0.75 km

16. Mr. Lee bought 4.5 pounds of apples at $1.79 per pound and 5.85 pounds of grapes at $1.40 per pound. Did he spend more money on the apples or grapes? (Round your answer to the nearest cent.)

3.11 Multiplying Decimals

Chapter 3 Review

LANGUAGE and CONCEPTS

Match each problem to the property it uses.

1. $3.1 \times 1 = 3.1$
2. $0 \times 0.05 = 0$
3. $20 \times 19 \times 5 = 19 \times 20 \times 5$
4. $46 \times 3 = (40 \times 3) + (6 \times 3)$
5. $(9 \times 1.5) \times 2 = 9 \times (1.5 \times 2)$
6. $4.5 \times 2 = 4.5 \times 2$

a. Commutative Property of Multiplication
b. Associative Property of Multiplication
c. Identity Property of Multiplication
d. Zero Property of Multiplication
e. Multiplication Property of Equality
f. Distributive Property

SKILLS and PROBLEM SOLVING

Use mental math to solve. (Section 3.2)

7. 60 × 700
8. 300 × 800
9. 9,000 × 60
10. 800 × 50
11. 8,000 × 5
12. 600 × 600
13. 400 × 900
14. 700 × 60

Estimate. (Section 3.3)

15. 74 × 4
16. $4.21 × 7
17. 110 × 92
18. 3,903 × 49

Use the Distributive Property to find the product. (Section 3.4)

19. 4×12
20. 6×25
21. 3×18
22. 8×32

Multiply. (Sections 3.5–3.7)

23. 67 × 30
24. 24 × 80
25. 65 × 42
26. 87 × 35
27. 213 × 97
28. 308 × 86
29. 509 × 70
30. 567 × 30

31. 456 × 800
32. 595 × 200
33. 789 × 806
34. 456 × 125

Multiply. (Section 3.9)

35. 8.3 × 10
36. 15.03 × 100
37. 0.068 × 10
38. 1.007 × 1,000
39. 203.005 × 1,000

Multiply. (Sections 3.10–3.11)

40. 85 × 1.2
41. 90 × 0.6
42. 25 × 2.3
43. 452 × 0.07
44. 4.1 × 0.9

45. 0.8 × 0.7
46. 0.1 × 0.1
47. 5.9 × 1.3
48. 2.7 × 5.3
49. 3.25 × 0.71

50. 4.56 × 0.34
51. 12.54 × 0.35
52. 10.05 × 0.03
53. 0.098 × 0.080

Solve by finding a pattern. (Section 3.8)

54. All the doors in the school are to be replaced. There are 50 rooms in the school. The principal needs to order metal digits to create the room numbers. The rooms are numbered in order from 1 to 50. How many of each digit does the principal need to buy? (*Hint:* Look for the patterns in the ones digits and the tens digits.)

55. Pedro has to make up a number pattern for math class. He thinks of the following sequence: 0.0054, 0.054, and 0.54. What is his pattern? What will be the next three numbers in his sequence?

Chapter 3 Review   87

Chapter 3 Test

Copy. Replace each ■ to make a true equation. Write the name of the property that you used.

1. 4 × ■ = 4
2. 56 × 8 = (■ × 8) + (■ × 8)
3. 18 × 2.3 = 18 × ■
4. 25 × 1.7 × 4 = 1.7 × ■ × 4

Multiply.

5. 200 × 90
6. 70 × 700
7. 6,000 × 80
8. 8,000 × 50

Estimate.

9. 456 × 34
10. 678 × 102
11. 895 × 6
12. 4,103 × 238

Multiply.

13. 87 × 40
14. 56 × 95
15. 546 × 23
16. 678 × 20

17. 457 × 600
18. 897 × 234
19. 2.3 × 10
20. 0.09 × 100

21. 8.007 × 1,000
22. 135.5 × 100
23. 55 × 0.9
24. 308 × 0.7

25. 1.3 × 0.8
26. 0.5 × 0.6
27. 134.7 × 2.5
28. 2.06 × 1.02

Solve.

29. A recipe for chocolate chip cookies calls for 0.75 cups sugar for one batch, 1.5 cups for two batches, and 2.25 for three batches. How many cups of sugar will be needed to make a dozen batches of cookies?

30. Margaret was buying oranges at the grocery store. She saw that oranges were $0.89 cents each or 3 for $2.50. Which is the better buy? Explain.

Change of Pace

Exponents

Some numbers can be written as the product of identical factors.

For example: $1,000 = 10 \times 10 \times 10$ or 10^3 ← exponent
base

▶ Notice the size and placement of the exponent.

The **exponent** 3 tells how many times the **base** 10 is used as a factor. 10^3 is a **power** of 10.

Write: 10^3
Say: ten to the third power (or ten cubed)

Some powers of 10 are shown below. What pattern do you notice?

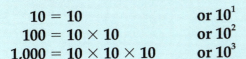

10^1 10^2 10^3

$10 = 10$	or 10^1	ten to the first power
$100 = 10 \times 10$	or 10^2	ten squared (or ten to the second power)
$1,000 = 10 \times 10 \times 10$	or 10^3	ten cubed (or ten to the third power)
$10,000 = 10 \times 10 \times 10 \times 10$	or 10^4	ten to the fourth power

Replace each ■ to make a true equation.

1. $10 \times 10 \times 10 = 10^■$
2. $10 \times 10 \times 10 \times 10 \times 10 = 10^■$
3. $100 = 10^■$
4. $100,000 = 10^■$
5. $1,000,000 = 10^■$
6. $400 = 4 \times 100$ or $4 \times 10^■$
7. $9,000 = 9 \times 1,000$ or $9 \times 10^■$
8. $300 = 3 \times 10^■$
9. $40 = 4 \times 10^■$
10. $5,000 = 5 \times 10^■$
11. $90,000 = 9 \times 10^■$
12. $8,000 = 8 \times 10^■$
13. $7,000,000 = 7 \times 10^■$

Write in standard form.

14. 2×10^3
15. 6×10^4
16. 8×10^2
17. 3×10^1
18. 5×10^6

Use powers of 10 to write numbers in expanded form.

$2,631 = 2,000 + 600 + 30 + 1$
$= (2 \times 1,000) + (6 \times 100) + (3 \times 10) + 1$
$= (2 \times 10^3) + (6 \times 10^2) + (3 \times 10^1) + 1$

19. 342
20. 3,492
21. 8,026
22. 38,904

Cumulative Test

1. What is the value of the 7 in 52,870,961?
   a. seven
   b. seventy
   c. seven thousand
   d. none of the above

2.   6
    + 7.2

   a. 7.8
   b. 7.26
   c. 13.2
   d. 1.32

3. 43 − 6.8 = _____
   a. 37.8
   b. 36.2
   c. 2.5
   d. 49.8

4.   3.6
   × 100

   a. 0.36
   b. 3.6
   c. 360
   d. 3,600

5. To win a prize, Clint must pick a number greater than 1.02 but less than 2.01. Which number must he choose to win?
   a. 0.999
   b. 1.015
   c. 1.623
   d. 2.015

6.   7,804
   ×   88

   a. 124,864
   b. 684,752
   c. 686,752
   d. none of the above

7. Joe and Tommy collect baseball cards. Joe has 7 fewer cards than Tommy. If together they have 83 cards, how many does Tommy have?
   a. 76
   b. 38
   c. 90
   d. 45

8. Mrs. Fields baked 8 trays of dinner rolls. Each tray held 24 rolls. How many dinner rolls did Mrs. Fields bake?
   a. 162 dinner rolls
   b. 192 dinner rolls
   c. 198 dinner rolls
   d. none of the above

9. The bar graph shows the number of times the vowels **A, E, I, O,** and **U** are used in this paragraph. How many more Es are used than Us?

   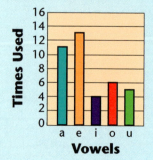

   a. 5
   b. 8
   c. 13
   d. none of the above

10. Sylvia walks for 1 hour 4 days a week. She rides her bicycle for 2 hours 3 days a week. How many hours a week does she ride her bicycle? Which of these facts is *not* needed?
    a. Sylvia rides her bicycle 3 days a week.
    b. Sylvia rides her bicycle for 2 hours.
    c. Sylvia walks 4 days a week.
    d. none of the above

CHAPTER 4

Division

Hilary Ross
Calvert Day School

4.1 Mental Math: Compatible Numbers

Objective: to use compatible numbers to estimate quotients

The fifth grade students are raising funds for their field trip by having a car wash. In 4 hours, they wash 126 cars. About how many cars did they wash each hour?

You can use **compatible numbers** to solve this problem. Compatible numbers are numbers that work well together (like compatible friends). Compatible numbers for 4 can be divided evenly by 4; 4, 8, 12, 16 and so on. They are often numbers from a basic fact.

Estimate 126 ÷ 4 using compatible numbers.

Step 1	Step 2
Decide on a compatible number. 120 is close to 126 and can be divided evenly by 4.	Divide. 30 $4\overline{)120}$ $\underline{-12}$ 0

The estimated quotient is 30. The fifth grade students washed about 30 cars each hour.

More Examples

A. Estimate. 6,034 ÷ 9
   Divide. 6,300 ÷ 9

700
$9\overline{)6,300}$

> Use the basic facts for 9 to help find compatible numbers.
>
> 9 × 6 = 54
> 9 × 7 = 63, use 63 because it is closer to 60.

The estimated quotient is 700. Since 6,300 is greater than 6,034, 700 is an overestimate.

B. Estimate. 568 ÷ 4
   Divide. 560 ÷ 4

140
$4\overline{)560}$

The estimated quotient is 140. Since 560 is less than 568, 140 is an underestimate.

TRY These

Choose the best pair of compatible numbers to use to make an estimate.

1. 3)146 a. 3 and 140 b. 3 and 150 c. 3 and 120
2. 8)876 a. 8 and 700 b. 8 and 880 c. 8 and 900
3. 5)492 a. 5 and 500 b. 5 and 400 c. 5 and 450
4. 7)6,587 a. 7 and 7,100 b. 7 and 6,500 c. 7 and 6,300

Exercises

Estimate the quotient. Tell whether your estimate is an *underestimate* or an *overestimate*.

1. 7)306
2. 2)104
3. 9)795
4. 3)367
5. 6)409
6. 5)655
7. 6)658
8. 4)238
9. 5)2,655
10. 6)5,234
11. 8)4,392
12. 2)2,567
13. 9)4,432
★ 14. 5)36,522
★ 15. 4)15,603
★ 16. 7)78,953

PROBLEM Solving

17. Estimate the number of yards in 1,450 feet. (1 yd = 3 ft)

Use the information below to answer questions 18–20.

3 tennis balls $3.80
4 soccer balls $98.32
6 baseballs $25.68

18. Estimate the cost of 1 tennis ball.
19. Estimate the cost of 1 soccer ball.
20. Estimate the cost of 1 baseball.

4.1 Mental Math: Compatible Numbers 93

Constructed Response

21. The school cafeteria used 55,967 paper napkins during the last year. If there are 9 months per school year, about how many napkins are used each month? Explain.

Compute.

22. 7.324 − 0.813	**23.** 12.06 − 8.953	**24.** $11.00 − 8.14	**25.** 8.4 − 2.02
26. 0.7 × 8	**27.** 0.91 × 4	**28.** 0.007 × 50	**29.** 0.037 × 100

MIND Builder

Operation Path

Follow the paths. Both paths end at the same number. What is it?

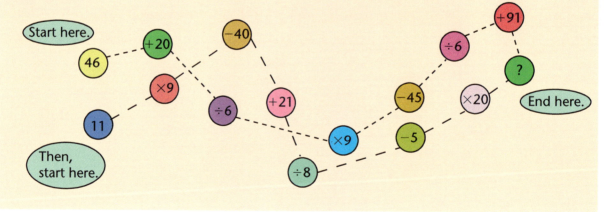

Problem Solving

Sum Fun

1. Arrange the eight number tiles into a square (with a hole in the center) so that the total number of dots on each side is 12.

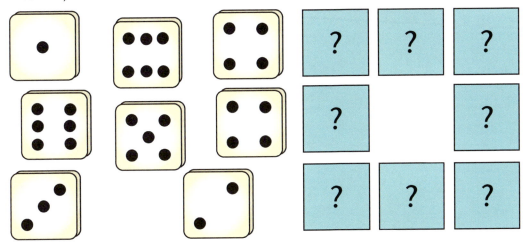

2. Now try to arrange these double number tiles into a square (with a hole in the center) so that the total number of dots on each side is the same.

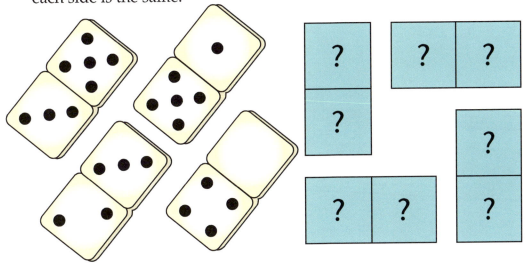

Extension

3. Make a number tile for each of the numbers 1–9. Arrange the tiles into a square so that the total number of dots in each row and each column is the same.

4.2 Zeros in the Quotient

Objective: to learn when to put zeros in the quotient

Inez Moore made 2 round trips to an antique store.

Her total mileage was 428 miles. How many miles was the antique store from her home?

$428 \div 4 = ?$ Estimate. $4\overline{)428} \rightarrow 4\overline{)400}$ with quotient 100

Step 1	Step 2	Step 3
Divide the hundreds.	Divide the tens.	Divide the ones.
$\begin{array}{r} 1 \\ 4\overline{)428} \\ \underline{-4} \\ 0 \end{array}$	$\begin{array}{r} 10 \\ 4\overline{)428} \\ \underline{-4} \\ 02 \end{array}$	$\begin{array}{r} 107 \\ 4\overline{)428} \\ \underline{-4} \\ 2 \\ \underline{-0} \\ 28 \\ \underline{-28} \\ 0 \end{array}$
	Since $2 < 4$, there are not enough tens to divide by 4. Write a zero in the quotient.	

It is 107 miles from Inez's home to the antique store.

Compare this answer with the estimate, 100.

More Examples

A. $\begin{array}{r} 60 \text{ R}5 \\ 8\overline{)485} \\ \underline{-48} \\ 05 \\ \underline{-0} \\ 5 \end{array}$

B. $\begin{array}{r} \$0.14 \\ 8\overline{)\$1.12} \\ \underline{-0} \\ 11 \\ \underline{-8} \\ 32 \\ \underline{-32} \\ 0 \end{array}$

C. $\begin{array}{r} 205 \text{ R}2 \\ 3\overline{)617} \\ \underline{-6} \\ 01 \\ \underline{-0} \\ 17 \\ \underline{-15} \\ 2 \end{array}$

D. $\begin{array}{r} 160 \text{ R}2 \\ 4\overline{)642} \\ \underline{-4} \\ 24 \\ \underline{-24} \\ 02 \\ \underline{-0} \\ 2 \end{array}$

Estimate; then divide. Check your answer.

1. 6)663
2. 5)$6.00
3. 8)851
4. 3)928
5. 2)611

Estimate; then divide. Check your answer.

1. 4)82
2. 3)61
3. 7)$7.14
4. 9)274
5. 6)363
6. 2)141
7. 5)527
8. 9)$9.36
9. 3)609
10. 3)$8.70
11. 7)842
12. 4)642
13. 5)2,515
14. 8)5,632
15. 4)3,628
16. 6)4,218

PROBLEM Solving

17. Alicia's family drove 816 miles in 4 days. On average, how far did they travel each day?

18. The distance between Baltimore, Maryland and Jackson, Mississippi is 1010 miles. If Mr. Hall wishes to make this drive in 2 days, how many miles should he drive each day?

19. Tickets to the school fair cost $6. The fair raised $6,730 the first day and $5,330 the second day. How many tickets were sold?

★ 20. Half of the sales from the fair in problem 19 went to purchase new gym equipment. The other half was given to the nearby SPCA. How much was given to the SPCA?

MIXED Review

Multiply.

21. 48 × 22
22. $81 × 81
23. 86 × 57
24. 739 × 214

4.2 Zeros in the Quotient 97

4.3 Dividing Greater Numbers

Objective: to divide with dividends greater than 1,000

The train shop has 1,450 feet of track. How many 6-foot lengths of track can be made? Divide 1,450 by 6.

Estimate. 1,200 is close to 1,450.

$$\begin{array}{r}200\\6\overline{)1{,}200}\end{array}$$

Step 1	Step 2	Step 3	Step 4
Divide the thousands.	Divide the hundreds.	Divide the tens.	Divide the ones.
$6\overline{)1{,}450}$ Are there enough thousands to divide by 6? No, but 14 can be divided by 6.	$\begin{array}{r}2\\6\overline{)1{,}450}\\-12\\\hline 2\end{array}$	$\begin{array}{r}24\\6\overline{)1{,}450}\\-12\\\hline 25\\-24\\\hline 1\end{array}$	$\begin{array}{r}241\ \text{R4}\\6\overline{)1{,}450}\\-12\\\hline 25\\-24\\\hline 10\\-\ 6\\\hline 4\end{array}$

There can be 241 6-foot lengths. There will be 4 feet of track left.

More Examples

A.
$$\begin{array}{r}532\ \text{R5}\\6\overline{)3{,}197}\\-30\\\hline 19\\-18\\\hline 17\\-12\\\hline 5\end{array}$$

B.
$$\begin{array}{r}\$22.91\\2\overline{)\$45.82}\\-4\\\hline 05\\-\ 4\\\hline 18\\-18\\\hline 0\ 2\\-\ 2\\\hline 0\end{array}$$

C.
$$\begin{array}{r}4{,}137\ \text{R8}\\9\overline{)37{,}241}\\-36\\\hline 12\\-\ 9\\\hline 34\\-27\\\hline 71\\-63\\\hline 8\end{array}$$

Estimate; then divide. Check your answer.

1. 2)1,960
2. 5)2,955
3. 3)9,840
4. 6)$84.30

Estimate; then divide. Check your answer.

1. 3)1,856
2. 8)9,207
3. 2)5,627
4. 3)$29.10
5. 9)$5,976
6. 6)1,272
7. 5)2,049
8. 3)6,841
9. 5)7,238
10. 2)5,216
11. 4)$7,736
12. 5)$63,145
13. 73,422 ÷ 6
14. 40,859 ÷ 4
15. 8,167 ÷ 5
16. 438,036 ÷ 8
17. 219,341 ÷ 9
18. 346,032 ÷ 5

PROBLEM Solving

19. Mr. Orta pays $8,136 for a used car. If he pays for the car in 9 months, how much does he pay each month?

20. A plane travels 2,070 miles in 6 hours. How far does it travel in 1 hour?

21. Four friends are going on a Caribbean cruise. The total cost is $2,768. If they share the cost evenly, how much will each friend pay?

Constructed Response

22. On average, Americans eat about 6 kg of chocolate per year. A fifth grade class figured out that they eat about 168 kg chocolate per year. How many students are in the class? Explain your reasoning.

4.3 Dividing Greater Numbers

4.4 Problem-Solving Application: Interpreting Remainders

Objective: to learn what to do with the remainder of a division problem

The director of a summer camp has many math problems to think about every day. Understanding the meaning of a remainder in division will help her.

Increasing the quotient

There are 204 children signed up for summer camp. Each counselor can have 8 children in her group. How many counselors will the director need to hire?

$$\begin{array}{r} 25\ R4 \\ 8\overline{)204} \\ -16 \\ \hline 44 \\ -40 \\ \hline 4 \end{array}$$

There will be 25 full groups of 8 children. Another group will be needed for the remaining 4 children. The director will need to hire 26 counselors.

Dropping the remainder

At the cookout, the campers made s'mores. They used 125 chocolate bars, which come in cases of 6. How many complete cases did they use?

$$\begin{array}{r} 20\ R5 \\ 6\overline{)125} \\ -12 \\ \hline 05 \end{array}$$

They used 20 complete cases. They only used 5 bars from case 21.

The remainder is the answer.

There are 32 children signed up for canoe lessons. Each instructor can teach a group of exactly 3 students. How many children will have to wait for others to sign up in order to have a full group?

$$\begin{array}{r} 10\ R2 \\ 3\overline{)32} \\ -3 \\ \hline 02 \end{array}$$

At any given time, 2 children will be waiting on the dock.

Changing the remainder to a fraction

The director purchased a large rope to use for a tug-of-war. The rope was 138 feet long. She cut the rope into 4 equal pieces, so the campers could be split into 4 groups. How long was each piece of rope?

$$\begin{array}{r} 34\frac{2}{4} \\ 4\overline{)138} \\ -12 \\ \hline 18 \\ -16 \\ \hline 2 \end{array}$$

▶ To write the remainder as a fraction, use the remainder as the numerator and the divisor as the denominator.

Each piece was $34\frac{2}{4}$ or $34\frac{1}{2}$ feet long.

Solve. Explain how you interpreted the remainder.

1. There are 54 dogs entered into a frisbee competition. They are divided as evenly as possible among the 5 different fields being used during the competition. How many dogs will be on most of the fields?
2. After initial elimination rounds, 6 dogs remained in the final round. The dogs caught 62 frisbees. On average, how many frisbees did each dog catch?
3. The 6 dogs ran a total of 39 miles during the entire competition. If each dog ran the same distance, how far did each dog run?
4. Shane ordered 112 new frisbees for the competition. They were evenly divided among 5 fields, with the extras being saved for the champion. How many frisbees did the champion receive?

Solve. Explain how you interpreted the remainder.

1. The school librarian has 235 new books to put on a bookshelf with 6 shelves. If she wants to put an equal number of books on each shelf, how many books will be on each shelf? How many books will be left over?
2. She purchased books equally from 3 different genres: science fiction, historical fiction, and mystery. On average, how many books did she purchase from each genre?

4.4 Problem-Solving Application: Interpreting Remainders

3. The remaining books were about snakes. How many books about snakes did she purchase?

4. She has a goal of reading all of the science fiction books over the summer. If she averages 4 books a week, how many weeks will she need to complete her goal?

Use the expression below for problems 5–7.

405 ÷ 8

5. Write and solve a word problem that requires you to increase the quotient.

6. Write and solve a word problem that requires you to drop the remainder.

7. Write and solve a word problem that requires you to use the remainder as the answer.

MIXED Review

Compute.

8. 94 − 36	9. 87 − 28	10. $7.08 − 2.56	11. 532 − 93	
12. 611 − 522	13. $74.29 − 43.65	14. 4,978 − 431	15. 6,071 − 5,862	

16. $3.86 − $1.07   17. 6,002 − 673

MIND Builder

Sums

Copy the triangle. Use the numbers 1–9. Place one number in each circle so that the sums along each side are the same. Use each number only once.

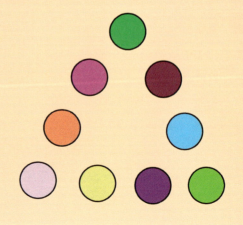

102   4.4 Problem-Solving Application: Interpreting Remainders

Cumulative Review

Write the value named by each underlined digit.

1. 3,4̲00,500
2. 7.1̲2
3. 36.92̲0
4. 432.961̲

Round to the underlined place-value position.

5. 47̲.9
6. 9̲,804
7. 27.8̲32
8. 65.34̲9

Find the sum or difference.

9. $25.69 + 11.98
10. $7.00 − 4.36
11. 96.35 + 12.00
12. 300.00 − 21.98

13. 2.611 + 43.09
14. 50.321 + 26.99
15. 23 + 81 + 1.02 + 0.99
16. 3.904 − 2.886
17. 213.61 − 5.290
18. 35.000 − 26.090

Multiply.

19. 262 × 8
20. 9,321 × 6
21. 42 × 42
22. 38 × 90
23. 5.89 × 17
24. 175 × 200
25. 7.86 × 12.3
26. 1.573 × 150

Estimate.

27. 92,611 + 5,008
28. $3.98 − 2.25
29. 286 × 7
30. 8)4,152

Divide.

31. 2)92
32. 4)876
33. 5)3,625
34. 6)2,366
35. 3)7,120
36. 9)8,181
37. 7)26,317
38. 8)48,024

Solve.

39. Student football tickets cost $1.25 each. What is the cost for 30 tickets?

4.5 Mental Math: Dividing by Multiples of 10

Objective: to use facts and patterns of zeros to divide by multiples of 10

At the campground, 80 canoes were rented for a total of 240 hours. What was the average number of hours each canoe was rented?

To find the average number of hours, divide 240 by 80. You can think of a related multiplication fact to help you divide.

80 × ■ = 240   8 tens × ■ = 24 tens
80 × 3 = 240   240 ÷ 80 = 3

Each canoe was rented for an average of 3 hours.

> Division undoes multiplication. You can show this with number families.

More Examples

A. 60, 2, 30
30 × 2 = 60   60 ÷ 2 = 30
2 × 30 = 60   60 ÷ 30 = 2

B. 540, 6, 90
90 × 6 = 540   540 ÷ 6 = 90
6 × 90 = 540   540 ÷ 90 = 6

C. 800, 40, 20
40 × 20 = 800   800 ÷ 20 = 40
20 × 40 = 800   800 ÷ 40 = 20

D. 3,500, 70, 50
70 × 50 = 3,500   3,500 ÷ 50 = 70
50 × 70 = 3,500   3,500 ÷ 70 = 50

TRY These

Complete the number family.

1. 40 × 2 = 80   80 ÷ 2 = 40
   2 × 40 = 80   80 ÷ 40 = ■

2. 90 × 7 = 630   630 ÷ 7 = 90
   7 × 90 = 630   630 ÷ 90 = ■

3. 60 × 4 = 240

4. 80 × 1 = 80

5. 30 × 9 = 270

6. 80 × ■ = 480

7. 30 × ■ = 600

8. 70 × ■ = 1,400

Copy and complete.

9. ■ × 30 = 90, so
   90 ÷ 30 = ■

10. ■ × 50 = 50, so
    50 ÷ 50 = ■

11. ■ × 20 = 800, so
    800 ÷ 20 = ■

12. ■ × 70 = 700, so
    700 ÷ 70 = ■

13. ■ × 90 = 7,200, so
    7,200 ÷ 90 = ■

14. ■ × 80 = 3,200, so
    3,200 ÷ 80 = ■

Exercises

Divide.

1. 80 ÷ 20   20 × ■ = 80
2. 360 ÷ 40   40 × ■ = 360
3. 420 ÷ 70
4. 240 ÷ 60
5. 480 ÷ 80
6. 200 ÷ 50
7. 560 ÷ 80
8. 700 ÷ 70
9. 1,600 ÷ 80
10. 3,600 ÷ 90
11. 6,400 ÷ 80
12. What is 6,300 divided by 70?
13. What is 3,000 divided by 60?

PROBLEM Solving

14. If 160 people rented 80 canoes, what was the average number of people in each canoe?

15. It cost $20.00 to rent each canoe. How many canoes can be rented for $1,600?

16. The boating center has 80 canoes that are green or red. If there is an equal number of each color, how many canoes of each color are there?

17. The boating center has 240 life jackets distributed evenly among 80 canoes. How many life jackets are in each canoe?

MID-CHAPTER Review

Estimate using compatible numbers. State whether your estimate is an *underestimate* or *overestimate*.

1. 4)467
2. 8)260
3. 5)119
4. 7)8,560

Divide.

5. 7)84
6. 6)185
7. 4)848
8. 4)524
9. 7)999
10. 9)2,309
11. 9)1,456
12. 3)45,892

4.5 Mental Math: Dividing by Multiples of 10

4.6 Dividing by Multiples of 10

Objective: to divide by multiples of 10

The Western Heritage Museum has 640 minutes of classic cowboy clips to be recorded on DVD. If each clip fills a 20-minute DVD, how many DVDs will be recorded?

Divide 640 by 20.

Step 1	Step 2	Step 3
Divide the hundreds.	Divide the tens.	Divide the ones.
3 $20\overline{)640}$ $6 < 20$ How many digits will be in the quotient?	3 $20\overline{)640}$ -60 4 **THINK** $20 \times 3 = 60$ ✓ $20 \times 4 = 80$ Compare 4 to the divisor, 20. 4 must be less than 20. Why?	32 $20\overline{)640}$ $-60↓$ 40 -40 0 $20 \times 1 = 20$ $20 \times 2 = 40$ ✓

They will record 32 DVDs.

More Examples

A. $10\text{ R}29$
$40\overline{)429}$
-40
29
-0
29

Check:
40 divisor
$\times10$ quotient
400
$+29$ remainder
429 dividend

B. $\$0.32$
$30\overline{)\$9.60}$
-90
60
-60
0

If there are no dollars, put a zero in the dollars column.

TRY These

Estimate; then divide. Check your answer.

1. 20)78
2. 80)728
3. 50)480
4. 60)517
5. 30)7,350
6. 40)8,337

Exercises

Estimate; then divide. Check your answer.

1. 20)620
2. 30)630
3. 20)620
4. 50)550
5. 60)682
6. 30)921
7. 60)747
8. 40)831
9. 70)3,150
10. 80)8,640
11. 40)7,440
12. 90)6,210
13. 80)9,234
14. 90)1,568
15. 50)1,786
16. 40)3,804

★ 17. 13,000 ÷ 40   ★ 18. 40,800 ÷ 50   ★ 19. 70 ÷ 5,841

PROBLEM Solving

Constructed Response

20. Mrs. Romero bought a used car that cost $8,580. If she makes payments of $90 twice a month, how many payments will she need to make in order to pay off the car? Explain how you interpreted the remainder.

21. See problem 20. Suppose Mrs. Romero pays only $30 twice a month. How many payments will she then need to pay off the car? How can you use your answer from problem 20 to check this answer?

4.6 Dividing by Multiples of 10

4.7 Two-Digit Divisors

Objective: to divide dividends with up to 3 digits by 2 digits

There are 275 guests at a bar mitzvah. They are to be seated at long tables of 24. How many tables are needed in order to accommodate all of the guests?

Divide. 275 ÷ 24   Estimate. 200 ÷ 20 = 10

Step 1	Step 2	Step 3

Divide the hundreds.
 	
24)275
 	
2 < 24; how many digits will be in the quotient?
	Divide the tens.
	
	   1
	24)275
	 −24↓
	   35
	
	THINK
	24 × 1 = 24 ✓
	24 × 2 = 48
		Divide the ones.
		
		   11 R11
		24)275
		 −24
		   35
		  −24
		   11

The quotient is 11 R11. Check this with the estimate to see if it is reasonable. This means that 12 tables will be needed. The quotient needs to be increased to seat the remaining 11 guests.

More Examples

```
      3 R3              8              4 R28
A. 31)96         B. 87)696      C. 56)252
   −93              −696           −224
     3                0              28
```

Check:
```
    56
  ×  4
   224
  + 28
   252
```

Estimate; then divide. Check your answer.

1. 31)96
2. 21)84
3. 29)70
4. 87)696
5. 56)252
6. 42)876

Estimate; then divide. Check your answer.

1. 19)95
2. 34)89
3. 22)110
4. 35)250
5. 59)118
6. 78)624
7. 42)126
8. 57)679
9. 36)380
10. 69)596
11. 21)567
12. 39)117
13. 23)997
14. 92)848
15. 81)490
16. 46)372
17. 34)276
18. 25)278
19. 32)702
20. 45)985

PROBLEM Solving

21. The Calvert football team has a budget of $864 for numbered jerseys. If there are 24 boys on the team, how much can be spent on each jersey?

22. The football team is going to a game that is 165 miles away. If their bus travels at an average speed of 55 mph, how many hours will it take them to get there?

23. There are 108 people going to the game from Calvert. If one bus holds 42 people, how many buses will the football team need to rent?

4.7 Two-Digit Divisors 109

4.8 Dividing 4- and 5-Digit Dividends

Objective: to divide dividends with up to 5 digits by 2 digits

There are 3,915 people to be seated at a dinner. Each table has seats for 16 people. To find the number of tables needed, divide 3,915 by 16.

Estimate.

$$16\overline{)3,915}^{\text{xxx}} \qquad 20\overline{)4,000}^{200}$$

There will be no thousands in the quotient.

Step 1	Step 2	Step 3
Divide the hundreds.	**Bring down the next digit from the dividend. Divide the tens.**	**Divide the ones.**
$\begin{array}{r} 2 \\ 16\overline{)3,915} \\ -3\,2 \\ \hline 7 \end{array}$ $16 \times 1 = 16$ $16 \times 2 = 32$ ✓ $16 \times 3 = 48$	$\begin{array}{r} 24 \\ 16\overline{)3,915} \\ -3\,2\downarrow \\ \hline 71 \\ -6\,4 \\ \hline 7 \end{array}$ $16 \times 3 = 48$ $16 \times 4 = 64$ ✓ $16 \times 5 = 80$	$\begin{array}{r} 244 \text{ R}11 \\ 16\overline{)3,915} \\ -3\,2 \\ \hline 71 \\ -6\,4\downarrow \\ \hline 75 \\ -6\,4 \\ \hline 11 \end{array}$
The first digit of the estimate is also the first digit of the quotient.		

If there are 244 tables, 11 people will not have a seat. So, 245 tables are needed.

Compared to the estimate, the answer seems reasonable.

More Examples

A. $\begin{array}{r} \$0.41 \\ 26\overline{)\$10.66} \\ -10\,4 \\ \hline 26 \\ -26 \\ \hline 0 \end{array}$

B. $\begin{array}{r} 164 \text{ R}20 \\ 22\overline{)3,628} \\ -2\,2 \\ \hline 1\,42 \\ -1\,32 \\ \hline 108 \\ -88 \\ \hline 20 \end{array}$

C. $\begin{array}{r} 924 \text{ R}11 \\ 46\overline{)42,515} \\ -41\,4 \\ \hline 1\,11 \\ -92 \\ \hline 195 \\ -184 \\ \hline 11 \end{array}$

Estimate; then divide. Check your answer.

1. 34)9,996
2. 16)3,915
3. 46)4,251
4. 72)8,424
5. 22)3,628
6. 61)4,598

Exercises

Estimate; then divide. Check your answer.

1. 91)2,093
2. 15)1,245
3. 39)1,530
4. 87)3,366
5. 53)4,028
6. 48)1,680
7. 64)2,944
8. 77)5,137
9. 74)59,734
10. 48)29,314
11. 31)13,600
12. 78)50,380
13. 54)26,136
14. 33)27,448
15. 56)10,040
16. 98)13,511

PROBLEM Solving

17. A cookie factory makes 34,000 cookies every month and packages them into boxes of 24. How many boxes are filled every month? How many cookies are left over?

18. The Blue Birds sold 14,448 boxes of cookies. If each Blue Bird sold 84 boxes of cookies, how many Blue Birds sold cookies?

19. In a canned food drive, 28 students collected 4,752 cans of food. What was the average number of cans collected by each person?

★ 20. How many more cans would have to be collected to raise the average to 175 per person?

21. How many dozen dinner rolls are needed to serve 156 people one roll each?
    a. 12
    b. 13
    c. 14
    d. 15

4.8 Dividing 4- and 5-Digit Dividends

4.9 Dividing Decimals by 10, 100, and 1,000

Objective: to divide by 10, 100, and 1,000

The fifth grade students are planning a walk-a-thon at their school to raise money for cancer research. The path they are planning is 22.5 kilometers long. If they want to evenly space out 10 water stations, how far apart will the water stations be?

Divide. 22.5 ÷ 10

When you divide by 10, you are making the number 10 times less.

22,500 ÷ 10 = 2,250
2,250 ÷ 10 = 225
225 ÷ 10 = 22.5
22.5 ÷ 10 = 2.25

> You can make a number 10 times smaller by moving the decimal point one place to the left.
>
> How do you think you could make a number 100 times smaller?
>
> 22.5 ÷ 100 = ?
>
> That's right, by moving the decimal point two places to the left.
>
> 22.5 ÷ 100 = 0.225

Examples

A. 45.6 ÷ 10 = 4.56
B. 45.6 ÷ 100 = 0.456
C. 45.6 ÷ 1,000 = 0.0456
D. 8 ÷ 10 = 0.8
E. 8 ÷ 100 = 0.08
F. 8 ÷ 1,000 = 0.008
G. 3.5 ÷ 10 = 0.35
H. 3.5 ÷ 100 = 0.035
I. 3.5 ÷ 1,000 = 0.0035

TRY These

Solve.

1. 95.6 ÷ 10
2. 7 ÷ 10
3. 1.8 ÷ 10
4. 50 ÷ 10
5. 95.6 ÷ 100
6. 7 ÷ 100
7. 1.8 ÷ 100
8. 50 ÷ 100
9. 95.6 ÷ 1,000
10. 7 ÷ 1,000
11. 1.8 ÷ 1,000
12. 50 ÷ 1,000

Exercises

Solve.
1. 340 ÷ 10
2. 13.7 ÷ 10
3. 9 ÷ 10
4. 0.4 ÷ 10
5. 293 ÷ 100
6. 40.9 ÷ 100
7. 12 ÷ 100
8. 0.5 ÷ 100
9. 135 ÷ 1,000
10. 95.3 ÷ 1,000
11. 2 ÷ 1,000
12. 0.3 ÷ 1,000

Compare using <, >, or =.
13. 35 ÷ 10 ● 330 ÷ 100
14. 2.4 ÷ 100 ● 0.2 ÷ 1,000
15. 135 ÷ 1,000 ● 3.5 ÷ 100
16. 0.5 ÷ 10 ● 5 ÷ 100
17. 11.8 ÷ 1,000 ● 0.08 ÷ 10
18. 1 ÷ 100 ● 10 ÷ 1,000

Solve for the missing number.
19. 6.7 ÷ ___ = 0.67
20. 58 ÷ ___ = 0.0058
21. 2.35 ÷ ___ = 0.0235
22. 157.9 ÷ ___ = 1.579
23. ___ ÷ 10 = 43.2
24. ___ ÷ 1,000 = 0.0098
25. ___ ÷ 100 = 14.0
26. ___ ÷ 100 = 0.0089

Problem Solving

27. A farmer owned a farm that was 1 square mile, or 640 acres. After his death, this farm was split evenly among his 10 sons. How many acres did each son inherit?

28. Oliver earned $5.00 from collecting 100 soda cans. How much did he get for each recycled can? How many cans would he have to collect to earn $50.00?

29. A dollar bill weighs about 0.033 ounce. How much would $100 worth of dollar bills weigh?

★ 30. Gordon weighs 100 pounds. How many dollar bills would he have to collect in order to equal his weight in dollar bills?

4.10 Converting Metric Units

Objective: to convert metric units

To change units in the metric system, it is helpful to remember that it is a base ten system.

Metric Units Equivalences

Length
10 millimeters (mm) = 1 centimeter (cm)
100 centimeters = 1 meter (m)
1,000 meters = 1 kilometer (km)

Capacity
1 liter (L) = 1,000 milliliters (mL)

Mass
1 kilogram (kg) = 1,000 grams (g)
1 metric ton (t) = 1,000 kilograms

To change measures follow these rules:

Larger unit ⟶ smaller unit, *multiply*.
Ex. 3 L = 3,000 mL (3 × 1,000 = 3,000)

Smaller unit ⟶ larger unit, *divide*.
Ex. 3,000 m = 3 km (3,000 ÷ 1,000 = 3)

When you multiply by a multiple of 10, move the decimal point right. When you divide by a multiple of 10, move the decimal point left.

More Examples

A. Change 3,456 millimeters to meters.

THINK 1 mm < 1 m, so divide.

1,000 mm = 1 m

3,456 ÷ 1,000 = 3.456   Move the decimal point three places left.
3,456 millimeters = 3.456 meters

B. Which is greater, 65 mm or 5.4 cm?

THINK 1 cm = 10 mm
       1 cm > 1 mm, so multiply.
Change centimeters to millimeters.

5.4 cm = 54 mm   Move the decimal point one place right.

65 mm > 54 mm so 65 mm > 5.4 cm.

TRY These

Decide if you *multiply* or *divide* to convert each unit.

1. kilograms to metric tons
2. centimeters to meters
3. milliliters to liters
4. grams to kilograms
5. kilometers to millimeters
6. liters to milliliters

Exercises

Complete.

1. 3.4 m = ___ mm
2. 634 mm = ___ cm
3. 12.5 cm = ___ mm
4. 8.71 km = ____ m
5. 3.2 m = ____ km
6. 12 mm = ____ m
7. 9 mm = _____ m
8. 14.85 m = _____ mm
9. 14 g = ____ kg
10. 3,100 g = ___ kg
11. 900 kg = _____ t
12. 456 kg = _____ g
13. 3.5 mL = _____ L
14. 8.5 L = _____ mL
15. 4.75 mL = ____ L

Compare using <, >, or =.

16. 87 mm ● 45 cm
17. 340 mm ● 3.5 m
18. 6,800 m ● 6.8 km
19. 375 m ● 3 km
20. 534 m ● 5.34 km
21. 5.25 kg ● 530 g
22. 0.2 t ● 2,000 g
23. 0.24 L ● 204 mL
24. 850 mL ● 8.5 L

PROBLEM Solving

25. Six 300-mL glasses are filled from a 2-L bottle of juice. How many milliliters are left in the bottle? How many liters are left?

Constructed Response

26. Grace says she weighs more than Chris. He weighs 40 kilograms and Grace weighs 40,000 grams. Is she correct? Explain.

4.10 Converting Metric Units 115

4.11 Dividing Decimals by Whole Numbers

Objective: to divide decimals by whole numbers

At a veterinarian's office, a vet weighed a litter of puppies. The combined weight was 14.8 kg. If there were 4 puppies, what was the average weight of each puppy?

To find the average weight of one puppy, you can divide the total weight by 4.

```
    3.7
4)14.8
 − 12
    2 8
  − 2 8
      0
```

- To divide a decimal by a whole number, first divide as if you were dividing whole numbers.
- Then, place the decimal point in the quotient directly above the decimal point in the dividend.

The average weight of each puppy is 3.7 kg.

Another Example

The vet spent 2.5 hours giving the 4 puppies their initial checkup. How much time was spent on each puppy?

```
   0.625
4)2.500
 − 2 4
     10
    − 8
     20
   − 20
      0
```

When you are working with decimals, avoid using remainders. You can do this by adding zeros to the dividend.

Remember that adding zeros to the end of a decimal does not change the value of the decimal.

The vet spent approximately 0.625 hours, or about 40 minutes, on each puppy.

TRY These

Divide. Do not write remainders. Add zeros if needed to solve.

1. 6)13.8
2. 7)24.5
3. 5)785.5
4. 8)2.44
5. 5)5.8
6. 3)$18.75
7. 2)8.1
8. 5)38

Exercises

Divide. Do not write remainders. Add zeros if needed to solve.

1. 4)8.3
2. 6)2.7
3. 9)77.4
4. 5)0.8
5. 5)14
6. 5)7.25
7. 6)$212.40
8. 25)8
9. 6)252.6
10. 4)1.1
11. 8)376.4
12. 3)$903.33
13. 4)0.23
14. 6)$2.28
15. 16)20
16. 15)3,071.25

PROBLEM Solving

17. A bag of 6 bagels costs $2.55. What is the cost of one bagel? Round your answer to the nearest cent.

18. Mr. Williams filled up his car with 22 gallons of gas. If the total cost was $67.98, what was the cost per gallon?

Constructed Response

19. Kevin can run at a pace of 12 mph. At this rate, how long will it take him to run a marathon that is 26.22 miles? Estimate to the nearest whole minute. Explain your reasoning.

4.11 Dividing Decimals by Whole Numbers

4.12 Problem-Solving Strategy: Eliminating Possibilities

Objective: to solve problems by eliminating possibilities

Simon Communications Inc. is awaiting letters from four of its top sales representatives. Their best salesperson lives in the United States and never drives through snow on the way to work. Who is their best salesperson?

1. READ   You need to know the name of the best salesperson. You have clues to help you.

2. PLAN   Draw rectangles to represent the letter from each salesperson.

| Brian Bell Minnesota | Jan Downes England | Bea Hall Hawaii | Chi Lee Japan |

3. SOLVE   You know that the best salesperson lives in the United States. The best salesperson does not live in England or Japan.

| Brian Bell Minnesota |  | Bea Hall Hawaii | |

You know that the best salesperson never drives through snow on the way to work. The best salesperson does *not* live in Minnesota.

|  |  | Bea Hall Hawaii |  |

You have eliminated all the possibilities except one.

The best salesperson must be from Hawaii, and her name is Bea Hall.

4. CHECK   Read the clues again. Bea Hall is from Hawaii, which is in the United States, and there is no snow on the streets of Hawaii.

TRY These

1. Claude's telephone bill is half as much as Bert's, and almost $10 less than Gretchen's. How much is Claude's bill?

 $36.40    $28.00    $18.20

Solve

1. Jenni and her cousins made paper-clip holders for birthday presents. The sides of Jenni's holder are all equal. Which holder did Jennie make?

   a. round   **b.** irregular shape   **c.** rectangular box   **d.** square box

2. Patrick found a U.S. coin. It had a building on one side and a man facing left on the other side. What coin did Patrick find?

MIND Builder

Estimation
If the figure below is uncurled, it would be 12 centimeters long.

Draw a figure of your own.

Begin drawing segments, and continue drawing until you think the length is 1 meter.

Match a string to your drawing and record the string's length. How close did you come to 1 meter?

Practice drawing other figures to improve your estimation skills.

4.12 Problem-Solving Strategy: Eliminating Possibilities

Chapter 4 Review

LANGUAGE and CONCEPTS

**Write *true* or *false*.
If false, write the sentence correctly.**

1. The answer in division is called the divisor.
2. Zero divided by any number is zero.
3. The dividend is the total amount to be divided.
4. If you multiply the quotient by the divisor and add the remainder, you will find the dividend.
5. Any number divided by 1 is 1.

SKILLS and PROBLEM SOLVING

Estimate using compatible numbers. (Section 4.1)

6. 8)1,785
7. 3)2,099
8. 6)1,050
9. 24)998
10. 39)2,879

Divide. (Sections 4.2–4.3)

11. 8)641
12. 5)605
13. 9)456
14. 4)210
15. 3)1,824
16. 8)3,468
17. 42,593 ÷ 7
18. 87,605 ÷ 4

Divide. (Sections 4.5–4.6)

19. 60)24,000
20. 800)3,200
21. $300 ÷ 5
22. 300)2,100
23. 40)3,421
24. 50)2,109
25. 60)8,500
26. $4,562 ÷ 30

Estimate; then divide. (Sections 4.7–4.8)

27. 29)345
28. 38)200
29. 56)988
30. 62)1,657
31. 97)3,456
32. 23)45,685

Find the equivalent value. (Sections 4.9–4.10)

33. 5.5 L = _____ mL **34.** 0.25 g = _____ kg **35.** 8.75 mL = _____ L

36. 893 mm = _____ cm **37.** 43.1 cm = _____ m **38.** 8.95 km = _____ m

Divide. Add zeros to the dividend, if needed. (Section 4.11)

39. $3.2 ÷ 8   **40.** 4 ÷ 5   **41.** 24.6 ÷ 3

42. $120.24 ÷ 4   **43.** 2.2 ÷ 5   **44.** $3.5 ÷ 2

Solve. (Sections 4.4, 4.11–4.12)

45. A group of 8 boys spent a total of $124 for baseball tickets and $66 for refreshments. What was each boy's share?

46. Patrick bought a dozen oatmeal cookies for $3.00. What is the price for 1 cookie?

47. Todd can put 62 baseball cards in a box. If he has 985 cards, how many boxes will he need to hold his collection?

48. If 1,588 eggs are to be packed in cartons holding 12 eggs each, how many extra eggs will there be after the cartons are packed? What fraction of another carton can be packed?

49. Tyler bought 3 trains for $11.94. What was the cost of 1 train?

50. If a kilowatt is 1,000 watts, how many kilowatts are in 4,218 watts?

Chapter 4 Test

Estimate using compatible numbers.

1. 2)118
2. 4)374
3. 7)3,029
4. 37)352
5. 91)4,501
6. 12)7,000

Divide.

7. 8)1,624
8. 7)$70.49
9. 7)83,417
10. 70)770
11. 80)829
12. 20)$50.74
13. 39)94
14. 41)7,462
15. 24)205.44
16. 2)$6.50
17. 4)14.4
18. 8)6

Compare using <, >, or =.

19. 835 mm ● 8.5 cm
20. 3.65 g ● 365 g
21. 2.3 km ● 2,300 m
22. 0.8 m ● 0.008 km
23. 1.2 kg ● 120 g
24. 35.4 mL ● 0.0354 L

Solve.

25. The 186 members of a fencing club are participating in a regional competition. If 22 fencers can fit into a room, how many rooms will be needed to accommodate all of the fencers?

26. The fencing club raised $191.25 by selling tickets to the competition. If they sold 45 tickets, what was the cost per ticket?

27. Mackenzie had a party. She spent $87.95 on the food and $37.65 on the decorations. If she had 16 guests, including herself, how much was spent per guest?

28. A ribbon 10.44 yards long is cut into 6 equal pieces. How long is each piece?

29. Ruth's message is in a balloon that is smaller than Tanya's. Brad's balloon is the same shape as Kate's. Which is Ruth's balloon?

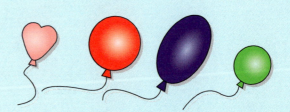

Change of Pace

Computation: Short Division

Dale Power's miniature race car completed 5 laps in 845 seconds. In how many seconds did it complete 1 lap?

Divide 845 by 5.   Estimate. 1,000 ÷ 5 = 200

You can use **short division** to divide. It is short because you compute mentally. You write only the quotient and the remainders as shown below.

Step 1	Step 2	Step 3
Divide the hundreds.	Divide the tens.	Divide the ones.
1 5)8₃45	1 6 5)8₃4₄5 ₃4 means 34 tens.	1 6 9 5)8₃4₄5 ₄5 means 45 ones.
THINK 5 × 1 = 5 8 − 5 = 3	**THINK** 5 × 6 = 30 34 − 30 = 4	**THINK** 5 × 9 = 45 The remainder is 0.

The car completed a lap in 169 seconds.

More Examples

A. 3)693 = 231

B. 8)25₁6 = 3 2

C. 6)6,4₄9₁5 = 1,0 8 2 R3

D. 4)$27.₃3₁6 = $6.84

Divide. Use short division.

1. 4)848
2. 3)936
3. 5)205
4. 6)612
5. 2)436
6. 3)642
7. 4)172
8. 5)115

Cumulative Test

1. Which replacement for ● makes this true?
   $8.28 + 92 \; ● \; 34.5 + 690$
   a. <
   b. >
   c. =
   d. none of the above

2. $6.27
   × 4
   a. $2.58
   b. $25.08
   c. $2,508
   d. none of the above

3. $50 \times 47 =$ _____
   a. 235
   b. 2,035
   c. 23,500
   d. none of the above

4. $60 \overline{)3{,}000}$
   a. 50
   b. 500
   c. 5,000
   d. none of the above

5. Timothy ran 0.62 kilometers. How many meters did he run?
   a. 6.2
   b. 62
   c. 620
   d. 6,200

6. $40 \overline{)8{,}320}$
   a. 208
   b. 280
   c. 2,080
   d. none of the above

7. Liam buys books for $18.00, $24.95, $6.95, and $13.00. He gives the clerk four $20 bills. How much is his change?
   a. $37.10
   b. $47.79
   c. $17.10
   d. $54.70

8. Emma ran the same distance each day for 10 days. If she ran 20 miles in all, how can you find the distance she ran each day?
   a. $20 + 10 = n$
   b. $20 \times 10 = n$
   c. $20 \div 10 = n$
   d. none of the above

9. There are 5 teams of fourth graders and 6 teams of fifth graders. Each team has 8 players. How many players are there in all?
   a. 19 players
   b. 50 players
   c. 88 players
   d. 240 players

10. Anna solves these problems:
    $1{,}294 - 798 =$ ■
    $36 \times 12 =$ ■
    $2{,}832 \div 6 =$ ■
    What are the answers from least to greatest?
    a. 424   462   496
    b. 432   462   494
    c. 432   472   496
    d. 496   432   472

CHAPTER 5

Fractions

Molly Hulsey
Missouri

5.1 Divisibility Rules

Objective: to determine when 2, 3, 5, 9, and 10 are factors of a number

Peter showed Ross six cards with different numbers written on each one.

| 590 | 690 | 543 | 990 | 695 | 582 |

He challenged Ross to guess which number was the one he had chosen. The only clue he gave is that his number is divisible by 2, 3, 5, 9, and 10. Which is Peter's number? To figure out Peter's challenge, Ross eliminated the numbers that are not divisible by 2, 3, 5, 9, and 10. One whole number is **divisible** by another whole number if the remainder is zero when you divide.

Divisible by 2? The number must be even.
590   690   ~~543~~   990   ~~695~~   582

Divisible by 5? The number must end in a 0 or a 5.
590   690   ~~543~~   990   ~~695~~   ~~582~~

Divisible by 10? The number must end with a 0.
All of the remaining numbers are divisible by 10.
590   690   ~~543~~   990   ~~695~~   ~~582~~

Divisible by 3? The sum of the digits must be divisible by 3.
590 ⟶ 5 + 9 + 0 = 14   14 ÷ 3 = 4 R2
690 ⟶ 6 + 9 + 0 = 15   15 ÷ 3 = 5
990 ⟶ 9 + 9 + 0 = 18   18 ÷ 3 = 6

~~590~~   690   ~~543~~   990   ~~695~~   ~~582~~

Divisible by 9? The sum of the digits must be divisible by 9.
690 ⟶ 6 + 9 + 0 = 15   15 ÷ 9 = 1 R6
990 ⟶ 9 + 9 + 0 = 18   18 ÷ 9 = 2

~~590~~   ~~690~~   ~~543~~   990   ~~695~~   ~~582~~

Peter's number was 990.

TRY These

Copy the table. Write either X or ✓ in the table to show whether each number is divisible by 2, 3, 5, 9, or 10.

	Number	Divisible by 2	Divisible by 3	Divisible by 5	Divisible by 9	Divisible by 10
1.	52					
2.	63					
3.	532					
4.	430					
5.	7,284					
6.	8,190					

Exercises

Use the rules to test each number for divisibility by 2, 3, 5, 9, and 10.

1. 837
2. 936
3. 850
4. 531
5. 1,349
6. 6,765
7. 2,682
8. 1,107
9. 94,654
10. 714,825
11. 234,190
12. 60,435

PROBLEM Solving

13. David says that it is never possible for a number to be divisible by both 2 and 5. William says it is always possible for a number to be divisible by both 2 and 5. Robert says it is sometimes possible for a number to be divisible by both 2 and 5. Who is right? Give *three* examples to support your answer.

14. Cole says that if a number is divisible by both 2 and 3, then it is also divisible by 6. Test out some numbers. Then explain whether or not you think this rule works.

15. There will be 118 people at a party. Each table can seat 5 people. Will all of the tables be full? Explain how you can tell without dividing.

Constructed Response

16. Can 3 people equally share 354 books? Explain how you can tell without dividing.

17. Write down your phone number, including the area code. Is it divisible by 2, 3, 5, 9, or 10? Explain.

5.1 Divisibility Rules

5.2 Prime and Composite Numbers

Objective: to identify numbers as prime or composite

Colleen has tomato plants that she wants to plant in her garden in equal rows.

She draws her plans by letting each square on the grid paper be 1 plant. She finds that 5 plants can be arranged in two ways.

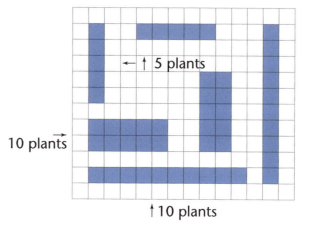

5 plants

10 plants

↑ 10 plants

$1 \times 5 = 5$ $5 \times 1 = 5$

Colleen finds that 10 plants can be planted in rows in more than two ways.

$2 \times 5 = 10$ $5 \times 2 = 10$ $1 \times 10 = 10$ $10 \times 1 = 10$

Exploration Exercise

1. Find all the row possibilities that can be made for each number from 1 to 12. Draw the shapes on grid paper as Colleen did above.

2. Record your data in a table.

Number	Combinations	Factors	Number of Factors
1	1 × 1	1	1
2	1 × 2, 2 × 1	1, 2	2
3	1 × 3, 3 × 1	1, 3	2
4	1 × 4, 4 × 1, 2 × 2	1, 2, 4	3

3. Circle the numbers that have only two factors.

4. Highlight the uncircled numbers.

5. Cross out the number **1**.

> ▶ The circled numbers are **prime numbers**. A prime number has exactly two different factors, itself and 1.
>
> The highlighted numbers are **composite numbers**. Composite numbers have more than two factors.
>
> One is neither prime nor composite. Why does it not fit into either category?

TRY These

List the factors of each number. Then identify whether each number is *prime* or *composite*.

1. 13   2. 24   3. 33   4. 51   5. 61   6. 16

Identify each number as *prime* or *composite*. Write each composite number as a product of two factors.

1. 2   2. 17   3. 15   4. 18   5. 21   6. 23
7. 41   8. 50   9. 43   10. 52   11. 37   12. 100

PROBLEM Solving

13. Can a prime number be a factor of a composite number? Why or why not?

14. Can a composite number be a factor of a prime number? Why or why not?

TEST Prep

15. Write 12 as a product of two factors.

    a. 3 × 5   b. 4 × 2   c. 3 × 4   d. 6 × 1

5.2 Prime and Composite Numbers

5.3 Prime Factorization

Objective: to write the prime factorization of a number

In this garden Emmett has planted peanuts in rows. There are 3 × 5 or 15 peanut plants. Is 15 a prime or composite number? What kind of numbers are 3 and 5?

Every composite number can be written as a product of prime factors. This is called **prime factorization.**

You can divide by prime numbers to find the prime factorization of a composite number.

The prime factorization of 15 is 3 × 5.

Example

Use a **factor tree** to find the prime factorization of 60.

▶ Circle the prime as soon as it appears.

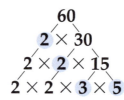

It does not matter what your starting factors are. The prime factorization will be the same.

```
     60
   6 × 10
 2×3 × 2×5
```

The prime factorization of 60 is **2 × 2 × 3 × 5**.
Factors should be written in order from least to greatest.

TRY These

Write the prime factorization of each composite number.

1. 9   2. 14   3. 12   4. 16   5. 27   6. 24

Exercises

Write the prime factorization of each composite number. Use factor trees.

1.
2.
3.
4.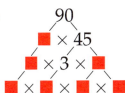

Write the prime factorization of each composite number.

5. 26
6. 50
7. 75
8. 12
9. 28
10. 105

11. 120
12. 54
13. 24
14. 81
15. 150
16. 210

17. 42
18. 36
19. 40
20. 72
21. 84
22. 96

23. What is the ones digit in any number that has 2 and 5 in its prime factorization?

24. Write the prime factorization of the numerator and the denominator of $\frac{25}{40}$. Are there any factors in common?

25. Which expression shows the prime factorization of 51?
    a. 7 × 7
    b. 5 × 11
    c. 3 × 7
    d. 3 × 17

26. A number less than 45 has three prime factors. Two of the factors are 3 and 7. What is the number?
    a. 36
    b. 42
    c. 63
    d. 41

5.3 Prime Factorization

5.4 Common Factors and Greatest Common Factor

Objective: to find the greatest common factor of pairs of numbers

Sam made 24 chocolate chip cookies and 40 peanut butter cookies to give as gifts. He wants to buy only one size box to pack the cookies. He wants the box to be as large as possible and to hold only one kind of cookie. How many cookies will each box hold?

Sam needs to find the **greatest common factor (GCF)** of 24 and 40. The greatest common factor is the product of the **common factors**.

First, use factor trees to find the prime factorization of 24 and 40.

$$
\begin{array}{c}
24 \\
2 \times 12 \\
2 \times 2 \times 6 \\
2 \times 2 \times 2 \times 3
\end{array}
\qquad
\begin{array}{c}
40 \\
4 \times 10 \\
2 \times 2 \times 2 \times 5
\end{array}
$$

Then, write the prime factorization for each number.

24: **2 × 2 × 2** × 3   Identify all the common factors between the two numbers.
40: **2 × 2 × 2** × 5

2 × 2 × 2 = 8, so the GCF of 24 and 40 is 8.

Sam should buy boxes large enough to hold 8 cookies.

He will have 3 boxes of chocolate chip cookies and 5 boxes of peanut butter cookies.

TRY These

Use factor trees to find the prime factorization of each number. Then write the GCF for each pair of numbers.

1. 4, 8
2. 12, 21
3. 6, 15
4. 10, 15

Exercises

Use factor trees to find the prime factorization of each number. Then write the GCF for each pair of numbers.

1. 18, 45
2. 11, 44
3. 13, 19
4. 32, 40
5. 30, 50
6. 16, 28
7. 32, 24
8. 45, 100
9. 16, 100
10. 28, 240
11. 35, 105
12. 20, 125

PROBLEM Solving

13. Alexa has 45 nonpareils and 60 chocolate kisses. What is the greatest number of cakes she can decorate if she must use all of the nonpareils and all of the chocolate kisses and each cake must have the same number of each?

★ 14. The members of the school's drama club are selling plants to raise money for their show. They have 12 marigolds, 24 petunias, 30 begonias, and 18 geraniums. They want to pack the plants in boxes that hold the same number. Each box will have only one kind of plant. What is the greatest number of plants each box should hold?

★ 15. The GCF of an odd number and an even number is 17. The greater number is 51. Find the other number.

Constructed Response

16. Carol says the GCF of 36 and 48 is 4. Explain Carol's mistake. Then state the correct answer.

5.4 Common Factors and Greatest Common Factor

5.5 Equivalent Fractions

Objective: to find equivalent fractions using shaded parts of figures and a number line

Rosa has two-fourths of a pizza. Ramon has the same amount of pizza cut into eighths. What other fraction names the amount of pizza Ramon has?

You can draw a model.

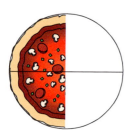 $\frac{2}{4} = \frac{4}{8}$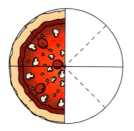

What do you notice about the number of pieces compared to the size of the pieces? Ramon has more pieces, but the pieces are smaller than Rosa's.

Ramon has four-eighths of a pizza.

Fractions like $\frac{2}{4}$ and $\frac{4}{8}$ that name the same amount are called **equivalent fractions**.

More Examples

A. Number lines can be used to show equivalent fractions.

Mark a number line to show sixths. Then fold it to make 3 equal pieces.

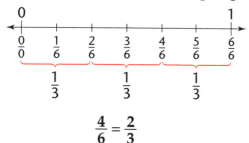

$$\frac{4}{6} = \frac{2}{3}$$

B. Counters can be used to show equivalent fractions.

Each counter is $\frac{1}{12}$ of the set.

Each group is $\frac{1}{4}$ of the set.

3 groups or $\frac{3}{4}$ of the set hold 9 counters or $\frac{9}{12}$ of the set.

$$\frac{3}{4} = \frac{9}{12}$$

TRY These

Write equivalent fractions to name the colored parts.

1.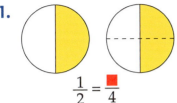
$\frac{1}{2} = \frac{\blacksquare}{4}$

2.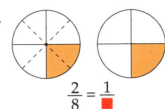
$\frac{2}{8} = \frac{1}{\blacksquare}$

3.
$\frac{\blacksquare}{3} = \frac{2}{\blacksquare}$

Exercises

Write two equivalent fractions to name the colored parts.

1.
2.
3.
4.
5.
6.
7.
8.

Use the number line to find a fraction equivalent to the given fraction.

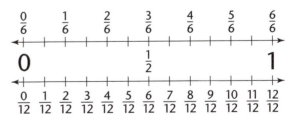

9. $\frac{1}{6}$
10. $\frac{8}{12}$
11. $\frac{4}{6}$
12. $\frac{5}{6}$
13. $\frac{1}{2}$
14. 1

Draw models to find equivalent fractions.

15. $\frac{6}{8} = \frac{\blacksquare}{4}$
16. $\frac{3}{5} = \frac{\blacksquare}{10}$
17. $\frac{2}{3} = \frac{\blacksquare}{12}$
18. $\frac{2}{10} = \frac{\blacksquare}{5}$

★ 19. $\frac{4}{12} = \frac{\blacksquare}{6} = \frac{\blacksquare}{3} = \frac{\blacksquare}{9}$

★ 20. $\frac{5}{10} = \frac{1}{\blacksquare} = \frac{3}{\blacksquare} = \frac{4}{\blacksquare}$

5.5 Equivalent Fractions 135

5.6 More Equivalent Fractions

Objective: to use multiplication and division to find equivalent fractions.

Mr. Perez has $\frac{1}{4}$ of a cake. He cuts that piece into 2 equal slices. Each slice is now $\frac{1}{8}$ of the cake, so Mr. Perez has $\frac{2}{8}$ of the cake.

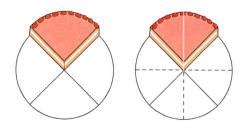

To find an equivalent fraction, you can multiply or divide the numerator and denominator of a fraction by the same number (not zero).

$\frac{1}{4} = \frac{1 \times 2}{4 \times 2} = \frac{2}{8}$ When you multiply the numerator, there are more pieces. When you multiply the denominator, the size of each piece is smaller.

Examples

You can use models to find equivalent fractions.

A. $\frac{6}{9} = \frac{\blacksquare}{3}$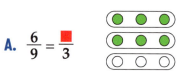

B. $\frac{2}{5} = \frac{4}{\blacksquare}$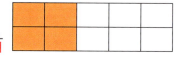

What did you do with 9 to make 3?

$\frac{6}{9} = \frac{6 \div 3}{9 \div 3} = \frac{2}{3}$ Notice when you divide, you get fewer, larger pieces.

What did you do to 2 to make 4?

$\frac{2}{5} = \frac{2 \times 2}{5 \times 2} = \frac{4}{\blacksquare}$

TRY These

Copy and complete.

1. $\frac{1}{3} = \frac{1 \times \blacksquare}{3 \times \blacksquare} = \frac{2}{6}$

2. $\frac{1}{4} = \frac{1 \times \blacksquare}{4 \times \blacksquare} = \frac{\blacksquare}{16}$

3. $\frac{6}{12} = \frac{6 \div \blacksquare}{12 \div \blacksquare} = \frac{1}{\blacksquare}$ ○○○○○○ / ○○○○○○

4. $\frac{6}{10} = \frac{6 \div \blacksquare}{10 \div \blacksquare} = \frac{\blacksquare}{5}$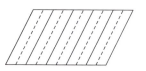

Exercises

Copy and complete. Draw a model if you would like.

1. $\frac{2}{5} = \frac{2 \times 2}{5 \times \blacksquare} = \frac{\blacksquare}{\blacksquare}$
2. $\frac{3}{4} = \frac{3 \times \blacksquare}{4 \times \blacksquare} = \frac{6}{\blacksquare}$
3. $\frac{1}{2} = \frac{1 \times \blacksquare}{2 \times \blacksquare} = \frac{\blacksquare}{10}$
4. $\frac{8}{10} = \frac{8 \div \blacksquare}{10 \div \blacksquare} = \frac{\blacksquare}{5}$
5. $\frac{6}{8} = \frac{6 \div \blacksquare}{8 \div \blacksquare} = \frac{\blacksquare}{4}$
6. $\frac{3}{12} = \frac{3 \div \blacksquare}{12 \div \blacksquare} = \frac{1}{\blacksquare}$

Replace each ■ with a number so that the fractions are equivalent.

7. $\frac{1}{2} = \frac{\blacksquare}{10}$
8. $\frac{1}{3} = \frac{5}{\blacksquare}$
9. $\frac{1}{4} = \frac{\blacksquare}{20}$
10. $\frac{1}{2} = \frac{\blacksquare}{8}$
11. $\frac{8}{16} = \frac{1}{\blacksquare}$
12. $\frac{8}{10} = \frac{4}{\blacksquare}$
13. $\frac{10}{12} = \frac{5}{\blacksquare}$
14. $\frac{12}{24} = \frac{1}{\blacksquare}$

Write the next three equivalent fractions.

15. $\frac{2}{3}, \frac{4}{6}, \frac{6}{9}, \blacksquare, \blacksquare, \blacksquare$
16. $\frac{3}{4}, \frac{6}{8}, \blacksquare, \blacksquare, \blacksquare$

Compare each pair of fractions. Are the fraction pairs equivalent? Write *yes* or *no*.

17. $\frac{5}{6}, \frac{10}{16}$
18. $\frac{6}{4}, \frac{3}{2}$
19. $\frac{30}{40}, \frac{3}{4}$
20. $\frac{2}{3}, \frac{14}{21}$

Problem Solving

21. Find a fraction equivalent to $\frac{1}{2}$. The numerator is a prime number. The denominator is a multiple of 5.

22. What is the fraction?
    - It is greater than $\frac{1}{2}$.
    - It is less than 1.
    - It is not equivalent to $\frac{1}{3}$.

    $\frac{1}{8} \quad \frac{6}{9} \quad \frac{9}{8} \quad \frac{3}{9}$

Mixed Review

Estimate.

23. 156 × 7
24. 39 × 15
25. $2.15 × 5
26. 347 × 24
27. 215 × 446

5.6 More Equivalent Fractions

5.7 Simplest Form

Objective: to find the simplest form of a fraction

Darlene used 6 of the 12 eggs in the box. What part of a dozen did she use?

You have learned to express $\frac{6}{12}$ in many ways using equivalent fractions. Find the **simplest form** of $\frac{6}{12}$.

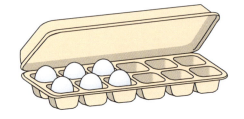

You can use fraction strips to find an equivalent fraction with the least number of pieces.

You can find the simplest form by dividing the numerator and denominator by their greatest common factor.

6: 2×3

12: $2 \times 2 \times 3$

GCF = 2×3 or 6

The simplest form of a fraction is the least number of like pieces.

Darlene used $\frac{1}{2}$ dozen eggs.

$\frac{6}{12} = \frac{6 \div 6}{12 \div 6} = \frac{1}{2}$

A fraction is in simplest form when both the numerator and the denominator can be divided by no other common factor than 1.

Example

Write $\frac{16}{20}$ in simplest form.

$16 = 2 \times 2 \times 2 \times 2$
$20 = 2 \times 2 \times 5$
GCF = 2×2 or 4

$\frac{16}{20} = \frac{16 \div 4}{20 \div 4} = \frac{4}{5}$

TRY These

Write each fraction in simplest form.

1.

   $\frac{3}{6} = \frac{\blacksquare}{\blacksquare}$

2.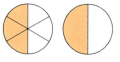

   $\frac{3}{9} = \frac{\blacksquare}{\blacksquare}$

3.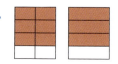

   $\frac{6}{8} = \frac{\blacksquare}{\blacksquare}$

4.

   $\frac{8}{12} = \frac{\blacksquare}{\blacksquare}$

Exercises

Write each fraction in simplest form. If the fraction is already in simplest form, just write the fraction.

1. $\frac{5}{10}$
2. $\frac{4}{6}$
3. $\frac{4}{8}$
4. $\frac{3}{12}$
5. $\frac{19}{38}$
6. $\frac{2}{10}$
7. $\frac{4}{12}$
8. $\frac{6}{21}$
9. $\frac{9}{12}$
10. $\frac{4}{7}$
11. $\frac{10}{12}$
12. $\frac{8}{10}$
13. $\frac{6}{18}$
14. $\frac{15}{21}$
15. $\frac{10}{15}$
16. $\frac{8}{16}$
17. $\frac{5}{8}$
18. $\frac{8}{24}$
19. $\frac{36}{50}$
20. $\frac{10}{100}$
21. $\frac{30}{60}$
22. $\frac{50}{100}$
23. $\frac{25}{100}$
24. $\frac{75}{100}$

Copy each number line. Then rewrite the number line so that all fractions are in simplest form.

25. Number line with marks: $\frac{0}{8}, \frac{1}{8}, \frac{2}{8}, \frac{3}{8}, \frac{4}{8}, \frac{5}{8}, \frac{6}{8}, \frac{7}{8}, \frac{8}{8}$

26. Number line with marks: $\frac{0}{10}, \frac{1}{10}, \frac{2}{10}, \frac{3}{10}, \frac{4}{10}, \frac{5}{10}, \frac{6}{10}, \frac{7}{10}, \frac{8}{10}, \frac{9}{10}, \frac{10}{10}$

★ **27.** Each notch on the ruler equals $\frac{1}{16}$ of an inch. Make a large number line to model an inch. Identify each fraction of an inch in simplest form on the number line and then find that notch on the ruler.

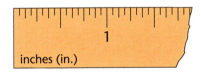

PROBLEM Solving

28. Robert earned $18 and spent $12. If he spends another $2, what fraction of his earnings does he still have?

29. In a survey, 16 out of 24 students preferred pizza for lunch. What part of the class did not prefer pizza? Express your answer in simplest form.

30. Tarik sold 18 of these cookies. What fraction of the cookies did he sell? Express your answer in simplest form.

★ **31.** What fraction of $4.00 is $0.75? Express your answer in simplest form.

5.7 Simplest Form

5.8 Problem-Solving Strategy: Choose the Method of Computation

Objective: to choose the correct method of computation

Martin's grandmother plans to travel across the country to visit her children and grandchildren. The map shows the routes and the distances she travels. How many miles will she travel?

1. READ — You need to find the number of miles she will travel. You know the routes and the distances between each city.

2. PLAN — The number of miles between each city is listed on the map. Find the number of miles between each city and add to get the total.

Which method of computation would you choose?

Write an equation. Let n represent what you are asked to find. Then solve using paper and pencil.

3. SOLVE

Minneapolis to San Francisco	2,151
San Francisco to Seattle	808
Seattle to Chicago	1,661
Chicago to Minneapolis	410

$n = 2{,}151 + 808 + 1{,}661 + 410$

$2{,}151 + 808 + 1{,}661 + 410 = 5{,}030$

4. CHECK — Check your calculations with an estimate.

$2{,}200 + 800 + 1{,}700 + 400 = 5{,}100$

Tell whether you would solve the problem by using *estimation, mental math,* or *paper and pencil.* Then solve.

1. The plane trip from Minneapolis to San Francisco took about 3 hours. About how many miles per hour did the plane travel?

2. Martin's grandmother makes the 5,030-mile trip 3 times a year. How many miles does she travel in 1 year?

1. Mr. Simpson works a 40-hour week for $8.60 an hour. What are his yearly earnings? (1 year = 52 weeks)

2. Imogene and her mother baked 10 dozen cookies. If they divided the cookies evenly among their 6 neighbors, how many cookies did each neighbor get?

3. Mr. Fernandez paid $39.50 for bowling shoes and $59.99 for a bowling ball. About how much did he spend on bowling equipment?

4. Sara Bishop bowled a perfect game of 300 pins. How many pins did she score in each of the 10 frames?

5. Bonnie paid $46.49 for a skirt and a pair of shoes. What was the cost of the skirt if the shoes cost $23.95 and the tax was $2.55?

6. Adele has $38 in cash and $42 in her savings account. Does she have enough money to buy a coat that costs $70?

7. Today is July 18. Jeremy's birthday is on November 4. How many days is it until Jeremy's birthday?

8. Jake uses 8 quarters to buy 24 gumballs. Don uses 20 dimes to buy gumballs that are 8¢ each. Who has made the better buy?

MID-CHAPTER Review

Write the prime factorization of each composite number.

1. 28
2. 50
3. 32
4. 105

Write each fraction in simplest form.

5. $\frac{3}{18}$
6. $\frac{12}{16}$
7. $\frac{15}{25}$
8. $\frac{34}{100}$

5.8 Problem-Solving Strategy: Choose the Method of Computation

5.9 Mixed Numbers and Improper Fractions

Objective: to understand the relationships between mixed numbers and improper fractions

After a pizza party, Mr. Robinson said that $1\frac{3}{4}$ pizzas were left. Mrs. Robinson said that $\frac{7}{4}$ pizzas were left. Who was correct?

You can think of this amount as either 1 or $\frac{4}{4}$.

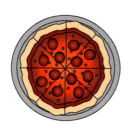 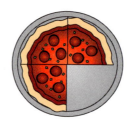

Both were correct. Mr. Robinson expressed the amount of pizza as a mixed number, while Mrs. Robinson used an improper fraction.

A **mixed number** is made up of a whole number and a fraction.

An **improper fraction** such as $\frac{7}{4}$ has a numerator that is greater than or equal to the denominator.

Another Example

Write a mixed number and an improper fraction to represent the colored area.

  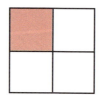

The mixed number is $2\frac{1}{4}$ and the improper fraction is $\frac{9}{4}$.

TRY These

Write each as a mixed number and as an improper fraction.

1.

2.

Exercises

Match each mixed number or improper fraction with the letter of the picture showing the value. The pictures may be matched to more than one number.

1. $3\frac{1}{3}$

2. $\frac{15}{8}$

3. 2

4. $2\frac{1}{2}$

5. $\frac{5}{2}$

6. $\frac{10}{9}$

7. $1\frac{1}{9}$

8. $\frac{10}{3}$

9. $\frac{10}{5}$

10. $1\frac{7}{8}$

a.

b.

c.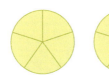

d.

e.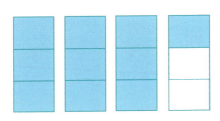

5.9 Mixed Numbers and Improper Fractions

Draw a picture or a diagram to help you solve.

11. Miguel has 15 eggs. Write what part of a dozen he has as a mixed number and as an improper fraction.

12. Huda's family ordered pizza for a family celebration. The picture shows how much pizza was eaten. Write the amount of pizza eaten as a mixed number and an improper fraction.

★ 13. If each person ate $\frac{2}{8}$ of a pizza, how many family members were at the celebration?

★ 14. The celebration started at 6:15 P.M. and ended at 9:30 P.M. Write how many hours it lasted as a mixed number and as an improper fraction.

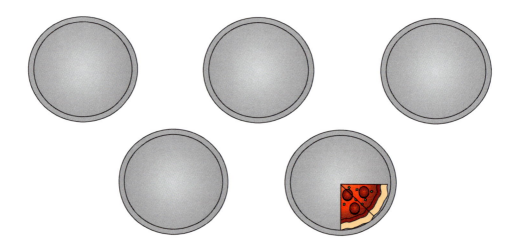

Cumulative Review

Write the value of each underlined digit.

1. 0.<u>3</u>0
2. 14,87<u>6</u>.05
3. 1.03<u>5</u>
4. 6,<u>8</u>97,458.1

Add or subtract.

5. 9 + 13.5
6. 42.1 − 0.165
7. 1,398.8 + 5.98
8. 3,090 − 193.5

Multiply.

9. 3.56 × 100
10. 0.56 × 10
11. 67 × 34
12. 876 × 125
13. 19 × 2.3
14. 0.45 × 0.9

Divide.

15. 488 ÷ 8
16. 17,543 ÷ 9
17. 546 ÷ 30
18. 1,245 ÷ 46
19. 4 ÷ 8
20. 20.5 ÷ 4

Solve.

21. Tom's car can travel 357 miles on 17 gallons of gasoline. How many miles per gallon does his car get?

22. A play started at 7:15 P.M. and ended at 9:43 P.M. How long did it last?

23. At a fabric store, a yard of fabric sells for $3.77. How much would 8 yards cost?

24. When Mr. Marks started driving on a trip, his car had 14.25 gallons of gas. At the end of his drive, his car had 9.6 gallons left. How many gallons did he use on the trip?

5.10 Renaming Improper Fractions as Mixed Numbers

Objective: to change improper numbers to mixed numbers

Mrs. Gardner has 7 sticks of butter. A box holds 4 sticks. Mrs. Gardner has $\frac{7}{4}$ (seven-fourths), or $1\frac{3}{4}$ (one and three-fourths), boxes.

To rename an improper fraction as a mixed number, divide the numerator by the denominator.

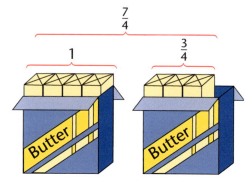

$$\frac{7}{4} \rightarrow 4\overline{)7}^{1}_{3}^{-4} \rightarrow 4\overline{)7}^{1\frac{3}{4}} \text{ boxes}$$

← This means 3 of the fourths are left.

Examples

A. $\frac{9}{3} = 3$ $3\overline{)9}$

B. $\frac{10}{6} = 1\frac{4}{6}$ or $1\frac{2}{3}$

C. $\frac{8}{5} = 1\frac{3}{5}$

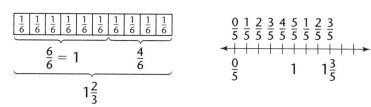

TRY These

Write *yes* if the fraction is improper and *no* if it is not. Rename each improper fraction as a whole number or mixed number in simplest form.

1. $\frac{6}{4}$ 2. $\frac{3}{2}$ 3. $\frac{5}{6}$ 4. $\frac{2}{1}$ 5. $\frac{9}{9}$ 6. $\frac{11}{12}$

Rename each improper fraction as a whole number or a mixed number in simplest form.

1. $\frac{21}{4}$
2. $\frac{17}{3}$
3. $\frac{15}{2}$
4. $\frac{25}{6}$
5. $\frac{7}{3}$
6. $\frac{19}{4}$
7. $\frac{15}{5}$
8. $\frac{18}{12}$
9. $\frac{18}{4}$
10. $\frac{22}{2}$
11. $\frac{13}{5}$
12. $\frac{9}{2}$
13. $\frac{50}{10}$
14. $\frac{20}{6}$
15. $\frac{48}{4}$
16. $\frac{30}{9}$
17. $\frac{68}{8}$
18. $\frac{45}{10}$

Write the letter for each number.

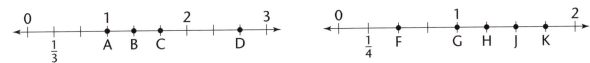

19. $1\frac{1}{3}$
20. $2\frac{2}{3}$
21. $\frac{5}{3}$
22. $1\frac{1}{2}$
23. $\frac{5}{4}$
24. $1\frac{3}{4}$

25. What is ten-fourths written as a mixed number in simplest form?

PROBLEM Solving

26. Mrs. Gardner used $\frac{8}{3}$ cups of flour to make coffee cake. Write that number as a mixed number in simplest form.

27. The cake needs 25 minutes to prepare, then bake for 35 minutes, and cool for 15 before it is cut. How many hours before the party does Mrs. Gardner need to begin the cake? Write your answer in simplest form.

28. A recipe calls for $\frac{5}{4}$ cups of milk. Write that number as a mixed number in simplest form.

Constructed Response

29. There are 7 pieces of cake left over from the party. If each piece is $\frac{1}{3}$ of a cake, in simplest form, how much cake is left over? Explain.

5.10 Renaming Improper Fractions as Mixed Numbers   147

5.11 Renaming Mixed Numbers as Improper Fractions

Objective: to rename mixed numbers as improper fractions

Tiara bought $3\frac{1}{2}$ yards of fabric to make handbags. If she needs $\frac{1}{2}$ yard for each handbag, how many handbags can she make?

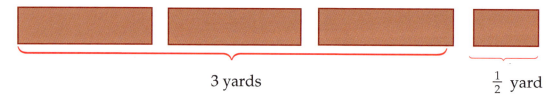

3 yards $\frac{1}{2}$ yard

Think of the 3 whole yards divided into halves. You can see that this will give you 6 halves.

When you rename mixed numbers as improper fractions, you multiply the whole number by the denominator of the fraction. This will divide the whole number into fractional parts.

$3\frac{1}{2}$ **THINK** $3 \times 2 = 6$ halves or $\frac{6}{2}$.

Then add this to the fraction that is there. $\frac{6}{2} + \frac{1}{2} = \frac{7}{2}$.

She can make 7 handbags.

Another Example

Rename $5\frac{1}{3}$ as an improper fraction.

THINK $5 \times 3 = 15$ thirds or $\frac{15}{3}$.

$\frac{15}{3} + \frac{1}{3} = \frac{16}{3}$

$5\frac{1}{3} = 3 \times 5 + 1 = \frac{16}{3}$ Keep the denominator.

TRY These

Rename each whole number as an improper fraction.

1. 2 as fourths
2. 3 as fifths
3. 5 as halves
4. 7 as thirds

Exercises

Rename each mixed number as an improper fraction.

1. $3\frac{1}{3}$
2. $5\frac{1}{4}$
3. $2\frac{2}{5}$
4. $5\frac{2}{3}$
5. $10\frac{1}{2}$
6. $9\frac{2}{5}$
7. $8\frac{3}{5}$
8. $4\frac{3}{4}$
9. $12\frac{2}{3}$
10. $1\frac{2}{7}$
11. $3\frac{5}{8}$
12. $5\frac{5}{6}$

PROBLEM Solving

13. How many beats are in $5\frac{3}{4}$ measures of music if there are 4 beats in each measure?

14. Noah ran $4\frac{1}{4}$ laps around the track. How many total fourths of the track did he run?

15. What number am I? I am a mixed number between 3 and 4. My value is greater than $\frac{30}{8}$. The denominator of my fractional part is 8.

16. What number am I? I am an improper fraction greater than 4 but less than $4\frac{1}{2}$. My denominator can be found by dividing 40 by 8. My numerator is an even number.

5.11 Renaming Mixed Numbers as Improper Fractions

5.12 Fractions as Decimals

Objective: to write fractions as decimals

Natalie's goal is to compete at the Olympics. She practices running on the $\frac{1}{2}$-mile lane to her house. How would that distance be shown on her pedometer?

The pedometer measures tenths of a mile.

One Way

- Write an equivalent fraction with a denominator of 10, or 100.

- Write the fraction as a decimal.

$\frac{1}{2} = \frac{5}{10}$ or 0.5   The pedometer would show 0.5 mile.

Examples

A. Write $\frac{1}{4}$ as a decimal.

$\frac{1}{4} \times \frac{25}{25} = \frac{25}{100} = 0.25$

B. Write $2\frac{2}{5}$ as a decimal.

The whole number stays the same. Change the fraction to tenths.

$\frac{2}{5} \times \frac{2}{2} = \frac{4}{10} = 0.4$

Include the whole number 2.4.

Another Way

C. A fraction can be changed to a decimal by dividing the numerator of the fraction by its denominator. To compute 1 ÷ 8 add a decimal point and zeros. The decimal point is placed directly above in the quotient.

$$\frac{1}{8} = \begin{array}{r} 0.125 \\ 8\overline{)1.000} \\ \underline{-8} \\ 20 \\ \underline{-16} \\ 40 \\ \underline{-40} \\ 0 \end{array}$$

TRY These

Copy and complete. Write each fraction as a decimal.

1. $\frac{3}{4} \times \frac{\blacksquare}{\blacksquare} = \frac{75}{100} = \blacksquare$

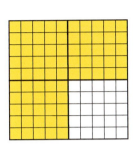

2. $\frac{7}{20} \times \frac{5}{5} \times \frac{\blacksquare}{\blacksquare} = \blacksquare$

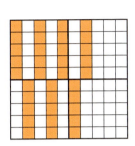

Exercises

Change each fraction to tenths. Then write the fraction or mixed number as a decimal.

1. $\frac{2}{5}$
2. $\frac{3}{5}$
3. $1\frac{4}{5}$
4. $\frac{1}{2}$
5. $5\frac{1}{2}$

Write an equivalent fraction with a denominator of 100. Then write each fraction as a decimal.

6. $\frac{1}{4}$
7. $\frac{9}{20}$
8. $9\frac{13}{25}$
9. $\frac{18}{50}$
10. $8\frac{12}{20}$

Write each fraction as a decimal.

11. $\frac{2}{8}$
12. $3\frac{1}{5}$
13. $\frac{3}{6}$
14. $\frac{15}{20}$
15. $15\frac{3}{8}$
16. $\frac{9}{12}$
17. $8\frac{4}{5}$
18. $4\frac{10}{20}$
19. $\frac{4}{16}$
20. $9\frac{1}{100}$

PROBLEM Solving

21. Write $3\frac{1}{4}$ as a decimal. In your own words, tell what method you used.

Constructed Response

★ 22. Write 0.375 as a fraction in simplest form. In your own words, explain your method.

5.12 Fractions as Decimals

Chapter 5 Review

LANGUAGE and CONCEPTS

Match.

1. The denominator of this fraction is 7.
2. The numerator of this fraction is 3.
3. This mixed number is less than 6.
4. This number is the simplest form of $\frac{18}{27}$.
5. This number is an improper fraction less than 2.
6. This is a prime number.
7. This is a composite number divisible by 10.
8. This is equivalent to $6\frac{3}{5}$.
9. This number is divisible by 3.
10. This number is divisible by 9.

a. 40
b. $\frac{15}{7}$
c. 6.6
d. $\frac{7}{5}$
e. 37
f. $\frac{3}{12}$
g. $\frac{2}{3}$
h. $5\frac{5}{6}$
i. 39
j. 45

SKILLS and PROBLEM SOLVING

Write the prime factorization of each number. State whether the number is *prime* or *composite*. (Sections 5.2–5.3)

11. 15
12. 42
13. 67
14. 75

Fill in the missing number to make a pair of equivalent fractions. (Section 5.4)

15. $\frac{4}{5} = \frac{\blacksquare}{15}$
16. $\frac{7}{8} = \frac{49}{\blacksquare}$
17. $\frac{24}{36} = \frac{2}{\blacksquare}$
18. $\frac{6}{21} = \frac{\blacksquare}{7}$

Write each fraction in simplest form. (Section 5.6)

19. $\frac{9}{12}$
20. $\frac{16}{20}$
21. $\frac{24}{48}$
22. $\frac{18}{35}$
23. $\frac{3}{18}$
24. $\frac{2}{7}$

Rename each improper fraction as a mixed number in simplest form. (Section 5.9)

25. $\dfrac{10}{4}$ 26. $\dfrac{23}{5}$ 27. $\dfrac{11}{9}$ 28. $\dfrac{14}{7}$ 29. $\dfrac{20}{7}$ 30. $\dfrac{18}{12}$

Rename each mixed number as an improper fraction. (Section 5.10)

31. $2\dfrac{1}{3}$ 32. $4\dfrac{2}{5}$ 33. $6\dfrac{3}{7}$ 34. $1\dfrac{2}{9}$ 35. $10\dfrac{1}{2}$ 36. $3\dfrac{3}{4}$

Write each fraction as a decimal. (Section 5.11)

37. $\dfrac{2}{10}$ 38. $\dfrac{4}{5}$ 39. $\dfrac{7}{8}$ 40. $\dfrac{14}{25}$ 41. $\dfrac{1}{8}$ 42. $\dfrac{17}{20}$

Solve. (Sections 5.7–5.9)

43. There are 12 chocolates per box. Mr. Lopez ordered 5 boxes for a party. When the party was over, 4 chocolates were left. Write a mixed number that represents the number of boxes of chocolates that were eaten.

44. Mr. Thrifty gave $\dfrac{1}{2}$ of his money to his daughter, Penny, and $\dfrac{1}{2}$ of what remained to her son, Bill. He donated the remaining $40,000 to the coin museum. How much money did Mr. Thrifty have?

45. At the farmer's market, the fruit stand has bags of apples and mangoes. There are a total of 64 apples and 48 mangoes. Each bag contains the same number of fruit. What is the greatest number of pieces of fruit each bag could contain?

Chapter 5 Review

Chapter 5 Test

Write the prime factorization of each number. State whether the number is *prime* or *composite*.

1. 24
2. 40
3. 57

Fill in the missing number to make a pair of equivalent fractions.

4. $\frac{9}{12} = \frac{\blacksquare}{36}$
5. $\frac{3}{7} = \frac{33}{\blacksquare}$
6. $\frac{40}{45} = \frac{8}{\blacksquare}$
7. $\frac{10}{35} = \frac{\blacksquare}{7}$

Rename each improper fraction as a mixed number is simplest form.

8. $\frac{12}{5}$
9. $\frac{42}{18}$
10. $\frac{32}{10}$
11. $\frac{20}{12}$

Rename each mixed number as an improper fraction.

12. $9\frac{1}{3}$
13. $6\frac{2}{5}$
14. $7\frac{3}{4}$
15. $3\frac{7}{8}$

Write each fraction as a decimal.

16. $\frac{9}{10}$
17. $\frac{3}{5}$
18. $\frac{5}{8}$
19. $\frac{2}{25}$

Solve. Write the answer as a mixed number if a part of the unit makes sense.

20. Fritz divides 15 cookies among 6 people. How many cookies does each person get?

21. A box contains 12 golf balls. How many boxes are needed to hold 54 golf balls?

Change of Pace

Tangram Activity

A tangram is a Chinese puzzle that is made from a square that is cut into pieces as shown. All seven pieces of the tangram together have a value of 1.

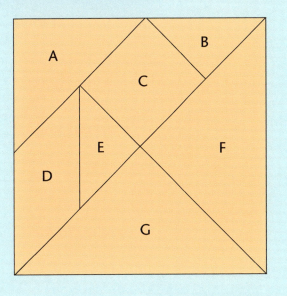

To find the value of each piece, complete these activities.

1. Trace the figure on another paper. Cut out the pieces and label them as shown.

2. Name the shape of each piece that is not a triangle.

3. How many of the pieces are triangles? Which triangles are equal in size?

4. Which three pieces are equal in size? (*Hint:* Use the smallest pair of congruent triangles to cover each area tested.)

5. Find two pieces that take up half of the tangram. What part of the whole square is triangle G? triangle F? Write the fraction on each triangle.

   $\frac{1}{2} = \frac{2}{\blacksquare}$

6. How many triangles the size of triangle A would fit on triangle G? What part of the whole square is triangle A? Write the fraction on triangle A.

   $\frac{1}{4} = \frac{2}{\blacksquare}$

7. Place triangles B and E on triangle A. What part of the whole square is triangle B? triangle E? Write the fraction on each triangle.

   $\frac{1}{8} = \frac{2}{\blacksquare}$

8. Use triangle B and E to find the value of C and D. Write the fraction on each piece.

9. Check your work by adding the values of all of the pieces of the puzzle. Is your answer 1?

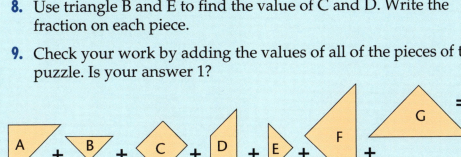

Cumulative Test

1.   92.93
    + 46.75

   a. 138.79
   b. 138.88
   c. 139.68
   d. 139.78

2. Which of the following is true?

   a. $\dfrac{2}{4} = \dfrac{3}{8}$ b. $\dfrac{4}{5} > \dfrac{2}{6}$

   c. $\dfrac{3}{8} < \dfrac{1}{8}$ d. $\dfrac{1}{5} > \dfrac{1}{2}$

3. Which is the correct decimal for six and twenty hundredths?

   a. 6.020
   b. 6.02
   c. 6.20
   d. 60.20

4. Which is the GCF of 20 and 25?

   a. 4
   b. 25
   c. 100
   d. 5

5. 37 − 20.382 = _____

   a. 16.618
   b. 17.608
   c. 17.618
   d. none of the above

6. What is 2.674 rounded to the nearest hundredth?

   a. 2.670
   b. 2.7
   c. 2.67
   d. 2.68

7. Rhonda has fit together 92 pieces out of a 1,000-piece jigsaw puzzle. What decimal indicates the amount of the puzzle she has fit together?

   a. 0.092
   b. 0.92
   c. 0.920
   d. 92

8. 6,240 ÷ 30 = _____

   a. 28
   b. 208
   c. 280
   d. 2,800

9. Charles rode his bicycle around a race course in 3.7 minutes during practice. In the actual race, he rode 0.47 minutes faster. What was his time in the race?

   a. 3.33 minutes
   b. 3.37 minutes
   c. 4.17 minutes
   d. none of the above

10. The product of 10.06 and 0.7 is _____.

    a. 70.42
    b. 7.42
    c. 7.042
    d. 7.24

CHAPTER 6

Adding and Subtracting Fractions

Ashley Gilmore
Maryland

6.1 Adding and Subtracting Like Denominators

Objective: to add and subtract fractions with like denominators

In Minya's apartment building there are curtains on $\frac{5}{8}$ of the windows and shades on $\frac{2}{8}$ of the windows. Add to find the total part having curtains and shades.

To add fractions with like denominators, add the numerators. Write the sum over the denominator.

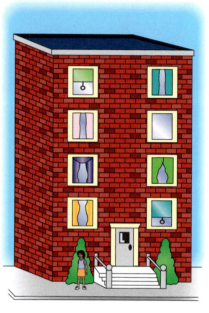

| $\frac{1}{8}$ | $\frac{1}{8}$ | $\frac{1}{8}$ | $\frac{1}{8}$ | $\frac{1}{8}$ | $\frac{1}{8}$ | $\frac{1}{8}$ | $\frac{1}{8}$ |

| $\frac{1}{8}$ | $\frac{1}{8}$ | $\frac{1}{8}$ | $\frac{1}{8}$ | $\frac{1}{8}$ | $\frac{1}{8}$ | $\frac{1}{8}$ |

> Fraction strips are models made from a unit strip divided into equal parts.

$$\frac{5}{8} + \frac{2}{8} = \frac{7}{8}$$

There are curtains or shades on $\frac{7}{8}$ of the windows.

Another Example

Minya had $\frac{5}{6}$ of a pizza to share. Her friends ate $\frac{4}{6}$. How much pizza is left?

To subtract fractions with like denominators, subtract the numerators. Write the difference over the denominator.

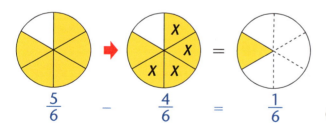

$$\frac{5}{6} - \frac{4}{6} = \frac{1}{6}$$

THINK
   5 sixths
− 4 sixths
   1 sixth

Only $\frac{1}{6}$ of the pizza is left.

TRY These

Write the addition or subtraction equation.

1. | $\frac{1}{10}$ | $\frac{1}{10}$ | $\frac{1}{10}$ | $\frac{1}{10}$ | $\frac{1}{10}$ | $\frac{1}{10}$ | $\frac{1}{10}$ | $\frac{1}{10}$ | $\frac{1}{10}$ | $\frac{1}{10}$ |

2.

3.

Exercises

Add.

1. $\frac{3}{8} + \frac{2}{8}$
2. $\frac{2}{5} + \frac{2}{5}$
3. $\frac{1}{6} + \frac{4}{6}$
4. $\frac{2}{10} + \frac{1}{10}$
5. $\frac{1}{9} + \frac{3}{9}$
6. $\frac{5}{8} + \frac{2}{8}$

Subtract.

7. $\frac{4}{8} - \frac{3}{8}$
8. $\frac{3}{6} - \frac{2}{6}$
9. $\frac{9}{10} - \frac{2}{10}$
10. $\frac{4}{5} - \frac{2}{5}$
11. $\frac{7}{9} - \frac{3}{9}$
12. $\frac{11}{12} - \frac{6}{12}$

Add or subtract.

13. $\frac{2}{6} + \frac{3}{6}$
14. $\frac{2}{10} + \frac{5}{10}$
15. $\frac{9}{10} - \frac{6}{10}$
16. $\frac{11}{12} - \frac{4}{12}$

Complete the operations in parentheses first.

★17. $\left(\frac{7}{10} + \frac{2}{10}\right) - \frac{6}{10}$

★18. $\left(\frac{8}{12} - \frac{3}{12}\right) + \frac{6}{12}$

★19. $\frac{4}{9} + \left(\frac{4}{9} - \frac{1}{9}\right)$

★20. What is $\frac{2}{9}$ more than $\frac{4}{9}$?

★21. What is the sum of $\frac{7}{10}$ and $\frac{1}{10}$?

PROBLEM Solving

22. Minya makes a fruit salad using $\frac{1}{4}$ cup melon, $\frac{1}{4}$ cup berries, and $\frac{2}{4}$ cup bananas. How many cups of fruit does she use?

23. Flight 402 completed a flight of 3,708 kilometers in 4 hours. If it flew the same distance each hour, how many kilometers did it fly each hour?

24. Henry has a piece of string $\frac{10}{12}$ of a foot long. If he cuts $\frac{3}{12}$ of a foot off, how long is the string now?

25. Jack walks $\frac{1}{8}$ of a mile to school and then $\frac{4}{8}$ of a mile to a friend's house. How far has he walked?

6.1 Adding and Subtracting Like Denominators

6.2 Answers in Simplest Form

Objective: to rename sums and differences in simplest form

The rain gauge shows that Center City had $\frac{2}{10}$ inch of rain before noon and $\frac{6}{10}$ inch after noon. How much rain fell that day?

Add. $\dfrac{2}{10} + \dfrac{6}{10} = \dfrac{8}{10}$

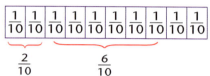

Write the answer in simplest form.

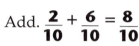

$\dfrac{8 \div 2}{10 \div 2} = \dfrac{4}{5}$

> The answer is in simplest form when the greatest factor that will divide into the numerator and denominator is 1.

In all, $\frac{4}{5}$ inch of rain fell that day.

More Examples

A. $\dfrac{8}{9} - \dfrac{2}{9} = \dfrac{6}{9}$ or $\dfrac{2}{3}$

B. $\dfrac{5}{6} + \dfrac{4}{6} = \dfrac{9}{6} = \dfrac{3}{2}$ or $1\dfrac{1}{2}$

C. $\dfrac{11}{8} - \dfrac{3}{8} = \dfrac{8}{8} = \dfrac{1}{1}$ or 1

TRY These

Is the answer in simplest form? Write *yes* or *no*.

1. $\dfrac{1}{8} + \dfrac{3}{8} = \dfrac{4}{8}$
2. $\dfrac{3}{6} - \dfrac{1}{6} = \dfrac{2}{6}$
3. $\dfrac{3}{4} + \dfrac{3}{4} = \dfrac{6}{4}$
4. $\dfrac{7}{5} - \dfrac{3}{5} = \dfrac{4}{5}$

Write each answer in simplest form.

5. $\dfrac{2}{3} + \dfrac{1}{3} = \dfrac{3}{3}$
6. $\dfrac{9}{10} - \dfrac{4}{10} = \dfrac{5}{10}$
7. $\dfrac{3}{8} + \dfrac{7}{8} = \dfrac{10}{8}$
8. $\dfrac{14}{9} - \dfrac{8}{9} = \dfrac{6}{9}$

160 6.2 Answers in Simplest Form

Exercises

Compute. Write the answers in simplest form. Remember to change any improper fraction to a mixed number.

1. $\frac{1}{6} + \frac{1}{6}$
2. $\frac{2}{4} + \frac{2}{4}$
3. $\frac{3}{8} - \frac{1}{8}$
4. $\frac{9}{10} - \frac{3}{10}$

5. $\frac{7}{9} + \frac{5}{9}$
6. $\frac{7}{10} + \frac{9}{10}$
7. $\frac{8}{6} - \frac{3}{6}$
8. $\frac{13}{12} - \frac{5}{12}$

9. $\frac{4}{5} + \frac{1}{5}$
10. $\frac{8}{9} + \frac{4}{9}$
11. $\frac{13}{10} - \frac{9}{10}$
12. $\frac{16}{12} - \frac{7}{12}$
13. $\frac{8}{10} + \frac{7}{10}$

Problem Solving

Write the answers in simplest form.

14. In Center City $\frac{3}{8}$ of the people work at home and $\frac{3}{8}$ of the people work away from home. What part of the people of Center City work?

15. Emilio draws a line $\frac{3}{4}$ inch long. He erases $\frac{1}{4}$ inch. How long is the line now?

Use the circle graph to answer questions 16–17.

16. How much of Kevin's time is spent in school and on homework and chores?

17. Name two things that together take one-half of Kevin's time.

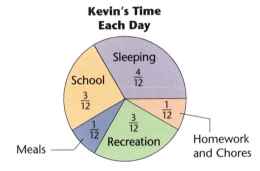

Kevin's Time Each Day

Mixed Review

Estimate.

18. 32 × 28
19. 61 × 78
20. 8 × 463
21. 4 × 921
22. 45 × 304
23. 16 × 769
24. 58 × 144
25. 98 × 923

6.2 Answers in Simplest Form

6.3 Least Common Denominators

Objective: to find the least common denominator for a pair of fractions

The fractions $\frac{1}{4}$ and $\frac{2}{4}$ have **common denominators**. The fractions $\frac{1}{3}$ and $\frac{1}{4}$ do not have common denominators. Fractions have common denominators when they have the same denominator.

Same denominators	Different denominators
$\frac{1}{4} \longleftrightarrow \frac{2}{4}$	$\frac{1}{3} \longleftrightarrow \frac{1}{4}$

To change $\frac{1}{3}$ and $\frac{1}{4}$ to fractions with common denominators think of the **multiples** of 3 and 4.

multiples of 3 → 3, 6, 9, ⑫, 15, 18, 21, ㉔, . . .

multiples of 4 → 4, 8, ⑫, 16, 20, ㉔, . . .

Two **common multiples** are 12 and 24. Always choose the smallest or **least common multiple**.

To change $\frac{1}{3}$ and $\frac{1}{4}$ to fractions with common denominators, change them both to equivalent fractions using the **least common denominator (LCD) of 12**.

$$\frac{1}{3} = \frac{\blacksquare}{12} \qquad \frac{1}{4} = \frac{\blacksquare}{12}$$

$$\frac{1}{3} \times \frac{4}{4} = \frac{4}{12} \qquad \frac{1}{4} \times \frac{3}{3} = \frac{3}{12}$$

$$\frac{4}{12} \longleftrightarrow \frac{3}{12} \quad \text{least common denominator}$$

TRY These

Find the LCD for these fractions.

1. $\frac{1}{2}, \frac{3}{4}$
2. $\frac{2}{6}, \frac{1}{2}$
3. $\frac{2}{5}, \frac{3}{4}$
4. $\frac{2}{3}, \frac{3}{8}$

Exercises

Change these pairs to fractions with least common denominators.

1. $\frac{1}{3}, \frac{4}{6}$
2. $\frac{1}{5}, \frac{4}{10}$
3. $\frac{1}{2}, \frac{7}{8}$
4. $\frac{1}{2}, \frac{1}{7}$
5. $\frac{3}{4}, \frac{2}{5}$
6. $\frac{1}{3}, \frac{1}{5}$
7. $\frac{4}{9}, \frac{2}{3}$
8. $\frac{3}{4}, \frac{1}{8}$
9. $\frac{2}{5}, \frac{4}{3}$
10. $\frac{1}{10}, \frac{4}{15}$
11. $\frac{2}{7}, \frac{1}{2}$
12. $\frac{1}{12}, \frac{5}{18}$

PROBLEM Solving

★ 13. Jack lives $\frac{1}{3}$ of a mile west of school and Jill lives $\frac{3}{4}$ of a mile east of school. How far is it from Jack's house to Jill's?

MIXED Review

14. If it is $\frac{1}{6}$ of a mile to Sally's house and Jane lives $\frac{3}{6}$ of a mile beyond Sally's house, how far is it to Jane's house?

15. What are the next three numbers in this pattern? 4, 8, 12, 16, 20

Solve.

16. $5{,}280 \div 60$
17. $3{,}254 \div 40$
18. $462 \div 66$
19. 426×18
20. 62×47

6.3 Least Common Denominators

6.4 Compare and Order Fractions

Objective: to compare and order fractions

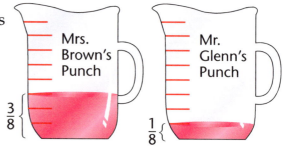

Who has more punch?

Since $\frac{3}{8}$ and $\frac{1}{8}$ have a common denominator, compare the numerators.

$$\frac{3}{8} > \frac{1}{8}$$

Mrs. Brown has more punch.

▶ When fractions have the same number as a denominator, they have a common denominator.

More Examples

Fractions that do not have common denominators can also be compared.

A. Use models to compare $\frac{3}{8}$ and $\frac{3}{4}$.

$\frac{1}{4}$		$\frac{1}{4}$		$\frac{1}{4}$		$\frac{1}{4}$	
$\frac{1}{8}$	$\frac{1}{8}$	$\frac{1}{8}$	$\frac{1}{8}$	$\frac{1}{8}$	$\frac{1}{8}$	$\frac{1}{8}$	$\frac{1}{8}$

▶ Since $\frac{1}{8}$ is less than $\frac{1}{4}$, 3 eighths would be less than 3 fourths.

$\frac{3}{8} < \frac{3}{4}$ You can also say $\frac{3}{4} > \frac{3}{8}$.

B. Compare $\frac{3}{4}$ and $\frac{5}{6}$.

You can use 12 as a common denominator to write equivalent fractions.

$\frac{3}{4} = \frac{9}{12}$ and $\frac{5}{6} = \frac{10}{12}$ ⟶ $\frac{9}{12} < \frac{10}{12}$, $\frac{3}{4} < \frac{5}{6}$

THINK What number can you use as a common denominator that is a multiple of both 4 and 6?
Multiples of 4: 4, 8, 12, 16
Multiples of 6: 6, 12, 18, 24

TRY These

Compare using <, >, or =.

1.

   $\frac{8}{8}$ ● $\frac{2}{2}$

2.

   $\frac{3}{4}$ ● $\frac{5}{8}$

3.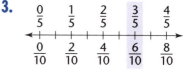

   $\frac{6}{10}$ ● $\frac{3}{5}$

164   6.4 Compare and Order Fractions

Exercises

Compare using <, >, or =.

1. $\frac{2}{8} \bullet \frac{1}{8}$
2. $\frac{6}{8} \bullet \frac{6}{8}$
3. $\frac{3}{12} \bullet \frac{4}{12}$
4. $\frac{6}{15} \bullet \frac{5}{15}$
5. $\frac{5}{5} \bullet \frac{2}{2}$
6. $\frac{2}{3} \bullet \frac{3}{4}$
7. $\frac{3}{4} \bullet \frac{9}{10}$
8. $\frac{7}{8} \bullet \frac{5}{6}$

★9. Compare $\frac{5}{10}$ and $\frac{45}{100}$.

★10. Compare $\frac{3}{100}$ and $\frac{3}{1,000}$.

Order from least to greatest.

11. $\frac{5}{6}$ $\frac{3}{6}$ $\frac{4}{6}$
12. $\frac{2}{3}$ $\frac{3}{4}$ $\frac{5}{12}$
★13. $\frac{5}{8}$ $\frac{5}{6}$ $\frac{5}{10}$

PROBLEM Solving

14. Jim takes $\frac{4}{5}$ of an hour to complete his homework and Tom takes $\frac{2}{3}$ of an hour. Who spends more time on homework?

15. At the theater $\frac{1}{2}$ of the people were children, $\frac{1}{5}$ were men, and $\frac{3}{10}$ were women. Order these groups from least to greatest.

MIND Builder

Mental Images

When folded, this pattern makes an open box.

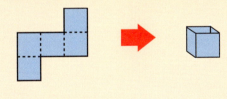

If the pattern is folded, will it make an open box? Write *yes* or *no*.

1.
2.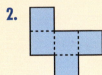
3.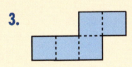

6.5 Adding Fractions with Unlike Denominators

Objective: to add fractions with unlike denominators

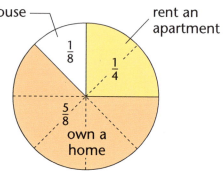
People Living in Greatown
rent a house — $\frac{1}{8}$
rent an apartment — $\frac{1}{4}$
own a home — $\frac{5}{8}$

To find the fractional part of the people who rent a place to live, add $\frac{1}{8}$ and $\frac{1}{4}$.

$\frac{1}{8} + \frac{1}{4} = ?$ Estimate. Since $\frac{1}{8}$ is less than $\frac{1}{4}$, $\frac{1}{8} + \frac{1}{4}$ must be less than $\frac{1}{2}$.

To add fractions that do not have the same denominator, rename the fractions using a common denominator.

Step 1	Step 2	Step 3
Find a common denominator. List multiples of the denominators. $\frac{1}{4}$ 4, ⑧, 12, 16, 24 $+\frac{1}{8}$ ⑧, 16, 24 Use 8 as the least common denominator.	Rename. $\frac{1}{4} = \frac{2}{8}$ $+\frac{1}{8} = \frac{1}{8}$	Add. $\frac{1}{4} = \frac{2}{8}$ $+\frac{1}{8} = \frac{1}{8}$ $\frac{3}{8}$

Note that 16 and 24 are also common denominators of 4 and 8, but using them means you have to simplify your answers.

In Greatown, $\frac{3}{8}$ of the people rent a place to live. $\frac{3}{8}$ is less than $\frac{1}{2}$. Compared to the estimate, this answer makes sense.

Rename each pair of fractions using a common denominator.

1. $\frac{1}{3}, \frac{1}{6}$
2. $\frac{5}{6}, \frac{1}{12}$
3. $\frac{4}{9}, \frac{2}{3}$
4. $\frac{1}{2}, \frac{3}{4}$
5. $\frac{1}{4}, \frac{7}{12}$
6. $\frac{1}{2}, \frac{7}{8}$

Exercises

Add. Write each sum in simplest form.

1. $\dfrac{1}{6} + \dfrac{1}{3}$
2. $\dfrac{1}{4} + \dfrac{3}{8}$
3. $\dfrac{1}{2} + \dfrac{1}{4}$
4. $\dfrac{2}{3} + \dfrac{1}{6}$
5. $\dfrac{3}{4} + \dfrac{1}{12}$
6. $\dfrac{5}{12} + \dfrac{1}{6}$

7. $\dfrac{5}{9} + \dfrac{1}{3}$
8. $\dfrac{2}{3} + \dfrac{7}{9}$
9. $\dfrac{3}{8} + \dfrac{1}{2}$
10. $\dfrac{2}{4} + \dfrac{1}{2}$

11. $\dfrac{5}{6} + \dfrac{1}{12}$
12. $\dfrac{2}{3} + \dfrac{1}{4}$
★13. $\dfrac{1}{4} + \dfrac{1}{2} + \dfrac{3}{4}$
★14. $\dfrac{2}{3} + \dfrac{1}{6} + \dfrac{1}{2}$

PROBLEM Solving

15. Davy drove $\dfrac{1}{3}$ of the way to Boston and then Henry drove $\dfrac{1}{2}$ of the way. How much of the trip have they driven?

16. Al waxed $\dfrac{1}{2}$ of the car before lunch. After lunch he waxed $\dfrac{2}{4}$ of the car. Has the entire car been waxed?

17. Roberto walks $\dfrac{5}{6}$ of a mile to the store and $\dfrac{1}{3}$ of a mile more to the theater. How far does he walk?

18. When Clifford went to visit Louis he walked $\dfrac{1}{8}$ of a mile. After visiting Louis he walked $\dfrac{2}{3}$ of a mile to Reid's house. How far did Clifford walk?

MIND Builder

Logical Reasoning

Follow the arrows to complete each square. Write the fractions in simplest form.

1.

	0	$\tfrac{1}{4}$	$\tfrac{1}{2}$	$\tfrac{3}{4}$
	$\tfrac{1}{2}$	$\tfrac{3}{4}$	?	?
	1	?	?	?
	$1\tfrac{1}{2}$	?	?	?

$+\tfrac{1}{2}$ ↑ $+\tfrac{1}{4}$ →

2.

	0	?	$\tfrac{1}{3}$	?
	?	?	?	?
	$\tfrac{2}{3}$	?	?	?
	?	?	?	?

$+\tfrac{1}{3}$ ↑ $+\tfrac{1}{6}$ →

6.5 Adding Fractions with Unlike Denominators

6.6 Subtracting Fractions with Unlike Denominators

Objective: to subtract fractions with unlike denominators

Mr. Gregory had $\frac{7}{8}$ cup of lemonade mix. He has used $\frac{1}{2}$ cup. How much of the mix is left?

$\frac{7}{8} - \frac{1}{2} = ?$ Estimate $\frac{7}{8} - \frac{1}{2}$ as $1 - \frac{1}{2}$ or $\frac{1}{2}$.

To subtract fractions with different denominators, find a common denominator.

Step 1	Step 2	Step 3
Find a common denominator and list multiples of the denominators. $\frac{7}{8}$ (8), 16, 24 $-\frac{1}{2}$ 2, 4, 6, (8), 10 Use 8 as the least common denominator.	Rename. $\frac{7}{8} = \frac{7}{8}$ $-\frac{1}{2} = \frac{4}{8}$	Subtract. $\frac{7}{8} = \frac{7}{8}$ $-\frac{1}{2} = \frac{4}{8}$ $\frac{3}{8}$

Use fraction strips to check your answer.

There is $\frac{3}{8}$ cup left.

$\frac{3}{8}$ is almost $\frac{1}{2}$. Compared to the estimate the answer makes sense.

TRY These

Rename each pair of fractions using the least common denominator.

1. $\frac{5}{8}, \frac{1}{2}$
2. $\frac{3}{4}, \frac{1}{8}$
3. $\frac{2}{3}, \frac{2}{9}$
4. $\frac{5}{6}, \frac{7}{12}$
5. $\frac{2}{5}, \frac{3}{10}$

Subtract. Write each difference in simplest form.

1. $\frac{7}{10} - \frac{1}{5}$
2. $\frac{3}{4} - \frac{1}{8}$
3. $\frac{5}{8} - \frac{1}{2}$
4. $\frac{3}{8} - \frac{1}{4}$
5. $\frac{1}{2} - \frac{1}{8}$

6. $\frac{8}{8} - \frac{3}{4}$
7. $\frac{6}{6} - \frac{2}{3}$
8. $\frac{3}{2} - \frac{3}{4}$
9. $\frac{10}{9} - \frac{1}{3}$
10. $\frac{7}{6} - \frac{5}{12}$

11. What is $\frac{2}{3}$ less than $\frac{8}{9}$?

12. How much greater is $\frac{3}{4}$ than $\frac{1}{2}$?

Copy. Replace each ■ to make a true equation.

13. $\frac{1}{4} + ■ = \frac{3}{4}$
14. $\frac{5}{8} + ■ = \frac{7}{8}$
15. $\frac{3}{12} + ■ = \frac{7}{12}$
16. $\frac{2}{3} + ■ = \frac{5}{6}$
17. $■ + \frac{3}{10} = \frac{4}{5}$
18. $■ + \frac{1}{6} = \frac{2}{3}$

PROBLEM Solving

19. The city plans to replace $\frac{4}{5}$ of the fence around the park by the end of the year. If $\frac{1}{3}$ has been done so far, how much more must be done to meet the goal?

20. Giles has $\frac{3}{4}$ of a tank of gas in his car. Driving to work and home takes $\frac{4}{7}$ of a tank. How much gas does he have remaining?

21. How much more is $\frac{5}{8}$ than $\frac{1}{6}$?

22. After the party, there was $\frac{5}{8}$ of a pizza left. John ate $\frac{1}{4}$ of a pizza for a snack. How much pizza remains?

MIXED Review

Estimate.

23. 29)630
24. 83)500
25. 17)3,743
26. 51)3,100
27. 12)445
28. 99)834
29. 68)5,087
30. 88)2,940

6.7 Adding and Subtracting Unlike Denominators

Objective: to add and subtract fractions with unlike denominators

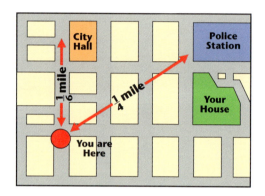

How much farther are you from the Police Station than from City Hall? To find the difference, subtract $\frac{1}{6}$ from $\frac{1}{4}$.

The denominators are not the same, so find a common denominator first. Then subtract.

Step 1	Step 2	Step 3
Find a common denominator. List multiples of the denominators. $\frac{1}{4}$ 4, 8, ⑫ $-\frac{1}{6}$ 6, ⑫, 18 Use 12 as the least common denominator.	Rename each fraction. $\frac{1}{4} = \frac{3}{12}$ $-\frac{1}{6} = \frac{2}{12}$	Subtract. $\frac{1}{4} = \frac{3}{12}$ $-\frac{1}{6} = \frac{2}{12}$ $\frac{1}{12}$

Check your answer with fraction circle models.

The Police Station is $\frac{1}{12}$ mile farther than City Hall.

Another Example

$\frac{2}{3} + \frac{1}{2} \rightarrow \frac{4}{6} + \frac{3}{6} = \frac{7}{6}$ or $1\frac{1}{6}$

3 : 3, ⑥

2 : 2, 4, ⑥

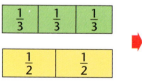
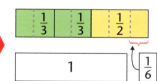

170   6.7 Adding and Subtracting Unlike Denominators

TRY These

Rename each pair of fractions using a common denominator. Then add or subtract.

1. $\frac{3}{4} + \frac{5}{6}$
2. $\frac{2}{3} + \frac{4}{9}$
3. $\frac{4}{5} - \frac{1}{2}$
4. $\frac{2}{3} - \frac{1}{4}$

Exercises

Add or subtract. Write the answer in simplest form.

1. $\frac{1}{2} + \frac{4}{5}$
2. $\frac{1}{6} + \frac{3}{4}$
3. $\frac{1}{3} + \frac{2}{3}$
4. $\frac{2}{5} - \frac{3}{10}$
5. $\frac{3}{5} - \frac{1}{2}$

6. $\frac{1}{4} + \frac{2}{3}$
7. $\frac{2}{9} + \frac{1}{6}$
8. $\frac{1}{6} - \frac{1}{8}$
9. $\frac{4}{9} - \frac{2}{6}$
★10. $\frac{1}{2} + \frac{5}{12} + \frac{1}{6}$

★11. $\frac{1}{4} + \frac{2}{3} + \frac{5}{6}$
★12. $\frac{3}{8} + \frac{1}{2} + \frac{1}{4}$

PROBLEM Solving

13. You walk $\frac{1}{2}$ of a mile to the police station and $\frac{2}{5}$ of a mile to the park. How far did you walk?

★14. A bottle contains $\frac{3}{4}$ of a liter of juice. Anna pours the juice into an empty glass until only $\frac{1}{12}$ of the juice is left in the bottle. If $\frac{1}{2}$ of the glass is filled with juice, what is the capacity of the glass?

★15. Arrange these fractions so that every side of the triangle has a sum of $1\frac{1}{2}$.

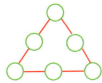
$\frac{3}{4}$ $\frac{1}{8}$ $\frac{1}{2}$ $\frac{1}{4}$ $\frac{5}{8}$ $\frac{3}{8}$

6.7 Adding and Subtracting Unlike Denominators

Problem Solving

Galactic Twins

After completing their training at the Galaxy Academy, twins Jacob and Sam Arwell are assigned as crew members on different starships. Each starship orbits around the same space station and stops to refuel after a certain number of orbits. Jacob's ship refuels after 5 orbits. Sam's ship refuels after 6 orbits.

If they leave at the same time, after which orbit will *both* Sam's and Jacob's ship be at the space station?

To solve this problem, make a table.

Refueling Stops

Jacob's ships	5	10	15			
Sam's ships	6	12	18			

Extension

Suppose another starship leaves at the same time and refuels after 4 orbits. When will the new ship meet Jacob's? Sam's? both Sam's and Jacob's? Draw a table to solve.

172   Problem Solving

Cumulative Review

Write the prime factorization of each number.

1. 18
2. 32
3. 43
4. 75
5. 80

Add or subtract.

6. 8,032 − 181
7. 79,384 − 46,532
8. 120,011 − 39,876
9. 24.1 + 14.56
10. $243.56 − $39
11. 300.1 − 0.876
12. 0.189 + 3.5
13. 0.098 − 0.006
14. 43 + 6.7 + 0.98

Multiply.

15. 25 × 30
16. 198 × 23
17. 167 × 51
18. 3.21 × 1.7
19. 85 × 0.09
20. 0.01 × 0.32

Divide. Express remainders as fractions.

21. 1,862 ÷ 3
22. 1,925 ÷ 40
23. 1,632 ÷ 5

Divide until the remainder is zero.

24. 1,862 ÷ 4
25. 6 ÷ 8
26. $10 ÷ 4

Rename mixed numbers as improper fractions and improper fractions as mixed numbers.

27. $4\frac{1}{3}$
28. $9\frac{2}{5}$
29. $7\frac{2}{3}$
30. $5\frac{4}{9}$
31. $\frac{15}{4}$
32. $\frac{12}{2}$
33. $\frac{21}{6}$
34. $\frac{14}{5}$

Solve.

35. Ling works at a conservatory and earns $360 in 20 hours. How much does she earn each hour? How much does she earn in 25 hours?

36. Ling spends $\frac{1}{4}$ of her day working in the Mediterranean room and $\frac{2}{3}$ of her day working in the Desert room. The remainder of her day is spent in the outside garden. What part of her day does she spend outside?

6.8 Problem-Solving Strategy: Relevant Information

Objective: to determine necessary information to solve problems

Not Enough Information

For the block party, Andre is making chicken salad. He mixes the $2\frac{3}{4}$ pounds of chicken with $1\frac{3}{8}$ pounds of grapes. He also mixes in pineapple, almonds, and salad dressing. What is the weight of the salad?

1. READ	You need to find the weight of the salad. You know the weight of the chicken and of the grapes.
2. PLAN	You can add the weight of each ingredient to find the total weight of the salad. However, you do not know the weight of the other ingredients. You do not have enough information to solve the problem.

Too Much Information

The block party started at 4:15. The DJ arrived at 5:30. By 11:00 that evening, everyone had gone home. How long was the party?

1. READ	You need to find the total length of the party. You know when the party started and when it ended. You do not need to know when the DJ arrived.
2. PLAN	Think about a clock to find the amount of time that passes between 4:15 and 11:00.
3. SOLVE	There are 45 minutes from 4:15 to 5:00 and 6 hours from 5:00 to 11:00. So the party lasted for 6 hours and 45 minutes.
4. CHECK	You used the starting time and ending time to find the length of the party.

TRY These

Write whether there is *not enough information* or *too much information*.

1. Elizabeth has 255 European stamps, 108 American stamps, and 98 Asian stamps. How many European and American stamps does she have?

2. Joe has some wood. He needs $1\frac{1}{4}$ feet to make a shelf for his room. How much wood will he have left?

Solve

Solve each problem. If there is not enough information, tell what information is needed. If there is too much information, state the extra facts.

1. James and Felicia are going to make spaghetti for a fundraiser dinner at their school. They have $50.00 to buy noodles, pasta sauce, cheese, garlic bread, and cookies. The garlic bread costs $7.25. Do they have enough money to buy the other ingredients?

2. At the supermarket they see 3 jars of sauce on sale for $7.99 and pasta on sale for $1.59 per pound. If they think that they will need 6 jars of sauce, how much will the sauce cost?

3. They buy a 32-ounce bag of cheese. James nibbles on 4 ounces of cheese on the way back home and eats 2 of the cookies. How much cheese is left?

4. They boil the water for the spaghetti. The box says that the spaghetti will take approximately 12–15 minutes to cook. When will the spaghetti be done?

MID-CHAPTER Review

Add or subtract. Write each answer in simplest form.

1. $\frac{3}{7} - \frac{1}{7}$
2. $\frac{5}{8} + \frac{3}{8}$
3. $\frac{3}{8} + \frac{12}{16}$
4. $\frac{7}{8} - \frac{2}{5}$

Compare using <, >, or =.

5. $\frac{5}{6} \bullet \frac{11}{12}$
6. $\frac{2}{7} \bullet \frac{4}{9}$
7. $\frac{1}{5} \bullet \frac{12}{7}$
8. $\frac{1}{2} \bullet \frac{20}{3}$

6.8 Problem-Solving Strategy: Relevant Information

6.9 Adding and Subtracting Mixed Numbers

Objective: to add and subtract mixed numbers with unlike denominators

Mrs. Wu used her computer for $2\frac{1}{2}$ hours this morning and $1\frac{1}{4}$ hours this afternoon. How long did she use her computer today? You can use models to show this addition.

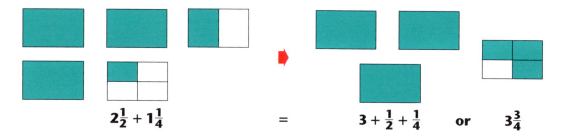

Mrs. Wu used her computer for $3\frac{3}{4}$ hours today.

Examples

A.
$$\begin{array}{r} 5\frac{11}{12} = 5\frac{11}{12} \\ +\ 4\frac{2}{3} = +\ 4\frac{8}{12} \\ \hline 9\frac{19}{12} = 10\frac{7}{12} \end{array}$$

Rewrite each mixed number using a common denominator for the fractions.

B.
$$\begin{array}{r} 10\frac{4}{6} = 10\frac{4}{6} \\ -\ 3\frac{1}{3} = -\ 3\frac{2}{6} \\ \hline 7\frac{2}{6} = 7\frac{1}{3} \end{array}$$

▶ Remember to keep the whole number when simplifying a mixed number.

TRY These

Add or subtract. Write each answer in simplest form.

1. $6\frac{3}{7}$
   $+\ 1\frac{2}{7}$

2. $2\frac{2}{9}$
   $+\ 9\frac{4}{9}$

3. $2\frac{3}{5}$
   $-\ 1\frac{1}{5}$

4. $5\frac{7}{8}$
   $-\ 3\frac{5}{8}$

Exercises

Add or subtract. Write each answer in simplest form.

1. $3 + 1\frac{1}{8}$

2. $2\frac{1}{6} + 1\frac{1}{3}$

3. $2\frac{7}{8} + 1\frac{7}{12}$

4. $1\frac{3}{5} + 3\frac{1}{2}$

5. $10\frac{2}{3} + 5\frac{4}{9}$

6. $3\frac{2}{3} - 2$

7. $2\frac{3}{5} - 2\frac{1}{10}$

8. $3\frac{5}{6} - 1\frac{1}{2}$

9. $4\frac{1}{2} - \frac{1}{6}$

10. $19\frac{7}{8} - 6\frac{3}{4}$

11. $6\frac{3}{8} + 7\frac{1}{2}$

12. $5\frac{1}{9} + 6\frac{2}{3}$

13. $8\frac{2}{3} - 3\frac{1}{2}$

14. $9\frac{5}{8} - 1\frac{1}{4}$

★15. $20\frac{1}{5} + 8\frac{1}{7}$

★16. $21\frac{5}{8} - 14\frac{1}{6}$

PROBLEM Solving

17. Suppose a line segment is $2\frac{3}{4}$ inches long. If you extend it by $1\frac{1}{8}$ inches, how long is the new line segment?

18. Suppose you draw a line segment $3\frac{5}{16}$ inches long. You then erase $2\frac{1}{4}$ inches. How long is the line segment now?

19. Mr. Wu ran out of gas. He has a can that holds $1\frac{1}{4}$ gallons and another one that holds $2\frac{3}{8}$ gallons. How much gas do both cans hold?

20. If $18\frac{5}{6}$ bushels of apples come from 1 tree, how many bushels come from 2 identical trees?

21. Martha caught 3 bass weighing $4\frac{1}{8}$ pounds, $6\frac{1}{2}$ pounds, and $3\frac{1}{4}$ pounds. What was the total weight of the fish?

22. What is the total weight of the oranges in the bags to the right?

$3\frac{3}{4}$ lb $4\frac{1}{8}$ lb

★23. The perimeter of the trapezoid is $36\frac{7}{8}$ inches. Side B is $1\frac{5}{8}$ inches longer than side A. How long is side A? How long is side B?

$7\frac{3}{8}$ in. $7\frac{3}{8}$ in.

A

B

6.9 Adding and Subtracting Mixed Numbers

6.10 Renaming to Subtract

Objective: to use renaming to subtract mixed numbers and fractions

At a soccer tournament, the first referee watched 4 games. The second referee watched $2\frac{2}{3}$ games before the third referee finished the rest of the games. How many more games did the first referee watch than the second referee?

Subtract. $4 - 2\frac{2}{3}$

Step 1	Step 2
Rename 4 as 3 + 1. Then rename 1 using 3 for the denominator. $4 = 3+1$ $= 3 + \frac{3}{3}$	Subtract the mixed numbers. $3\frac{3}{3}$ $-2\frac{2}{3}$ $\overline{1\frac{1}{3}}$

The first referee watched $1\frac{1}{3}$ games more than the second referee.

You can also rename a mixed number.

Subtract. $7\frac{1}{4} - 2\frac{3}{4}$

Step 1	Step 2
Rename $7\frac{1}{4}$. $7\frac{1}{4} = 7 + \frac{1}{4}$ $= 6 + 1 + \frac{1}{4}$ $= 6 + \frac{4}{4} + \frac{1}{4}$ $= 6 + \frac{5}{4}$	Subtract the mixed numbers. $6\frac{5}{4}$ $-2\frac{3}{4}$ $\overline{4\frac{2}{4}} = 4\frac{1}{2}$

TRY These

Write each difference in simplest form.

1. $5 - 1\frac{1}{4}$
2. $10 - 3\frac{2}{3}$
3. $9\frac{1}{5} - 6\frac{4}{5}$
4. $2\frac{1}{4} - \frac{3}{4}$
5. $8\frac{1}{8} - 3\frac{5}{8}$
6. $11\frac{1}{6} - 8\frac{4}{6}$

Exercises

Write each difference in simplest form.

1. $4\frac{3}{8} - 3\frac{5}{8}$
2. $5 - 2\frac{2}{3}$
3. $6\frac{1}{2} - 1\frac{3}{5}$
4. $3\frac{1}{4} - 1\frac{1}{2}$
5. $5 - 3\frac{1}{6}$
6. $4\frac{1}{8} - \frac{7}{8}$
7. $25 - 11\frac{2}{9}$
8. $30\frac{1}{8} - 19\frac{5}{16}$
9. $10\frac{5}{9} - 3\frac{7}{9}$
10. $112\frac{1}{2} - 89\frac{3}{16}$
11. $21\frac{2}{5} - 13\frac{8}{10}$
12. $8\frac{1}{4} - 6\frac{5}{6}$

PROBLEM Solving

13. Maddie lives $3\frac{2}{5}$ miles from her school and $5\frac{1}{4}$ miles from her dad's office. How much closer does she live to school than to her dad's office?

15. The longest snake ever recorded was a reticulated python that was almost 33 feet long. The shortest snake is the thread snake, which is barely $4\frac{1}{4}$ inches long. What is the difference in length between the two snakes?

★14. A piano teacher kept track of how much her students practiced during the week. Adrian practiced for $8\frac{1}{4}$ hours, Alexi practiced for $4\frac{5}{6}$ hours, and Walter practiced for $4\frac{11}{12}$ hours. What is the greatest difference between the times the students practiced?

6.10 Renaming to Subtract

6.11 Customary Measurement

Objective: to convert between customary units

Length

Carrie is climbing a rope. She has already climbed 40 inches. How many feet has she climbed?

Rename 40 inches as feet.

THINK
1 foot (ft) = 12 inches (in.)
1 yard (yd) = 3 ft or 36 in.
1 mile (mi) = 5,280 ft or 1,760 yd

40 in. = ☐ ft

$\frac{40}{12} = 3\frac{4}{12} = 3\frac{1}{3}$

40 inches = $3\frac{1}{3}$ feet

> To change measures follow these rules.
> To go from a large unit ⟶ small unit, *multiply*.
> To go from a small unit ⟶ large unit, *divide*.

Weight

Rename $7\frac{1}{2}$ pounds as ounces.

THINK
1 pound (lb) = 16 ounces (oz)
1 ton (T) = 2,000 lb

$7\frac{1}{2}$ lb = ☐ oz

7.5 × 16 = 120   Change the fraction to a decimal to multiply.

$7\frac{1}{2}$ lb = 120 oz

Capacity

Rename 30 cups as quarts.

THINK
1 cup (c) = 8 fluid ounces (fl oz)
1 pint (pt) = 2 c
1 quart (qt) = 2 pt or 4 c
1 gallon (gal) = 4 qt

30 c = ☐ qt

30 ÷ 4 = $7\frac{1}{2}$ qt

There are 4 cups in 1 quart, so divide 30 by 4.

Decide if you *multiply* or *divide* to rename each unit.

1. gallons to cups
2. ounces to pounds
3. feet to inches
4. tons to ounces
5. inches to yards
6. cups to quarts

Complete.

1. $3\frac{1}{2}$ ft = ☐ in.
2. $5\frac{1}{4}$ lb = ☐ oz
3. 12 c = ☐ qt
4. 50 in. = ☐ ft
5. 15 gal = ☐ qt
6. 128 oz = ☐ lb
7. 90 yd = ☐ ft
8. $3\frac{1}{2}$ qt = ☐ pt
9. $4\frac{1}{2}$ T = ☐ lb
10. $3\frac{1}{3}$ yd = ☐ ft
11. $\frac{1}{2}$ gal = ☐ qt
★12. $15\frac{1}{2}$ ft = ☐ yd

PROBLEM Solving

13. Maria uses 8 inches of ribbon to make one bookmark. If she makes 20 bookmarks, how many feet of ribbon does she use?

14. The average person drinks 1 pint of milk a day. At this rate, how many gallons will a person drink in a year?

15. Maggie bought $2\frac{1}{2}$ pounds of coleslaw for a picnic. How many ounces of coleslaw were eaten if there are $\frac{7}{8}$ ounces left?

16. A marathon is $26\frac{1}{5}$ miles long. How many feet is that?

17. Fido weighs $75\frac{1}{2}$ pounds and Fifi weighs 130 ounces. How many pounds do they weigh together?

Constructed Response

18. A hockey field is 3,600 inches long. A soccer field is 330 feet long and a football field is 120 yards long. Which sport has the longest playing field? Explain.

6.11 Customary Measurement

6.12 Computing with Measurement

Objective: to add and subtract customary and metric units

Ricky keeps a record of the amount of time he spends each week exercising. How much did Ricky exercise during his first 3 weeks?

Week	Hours	Minutes
1	3	12
2	2	45
3	3	50
4	3	30

To find the total, you add.

```
  3 h  12 min
  2 h  45 min
+ 3 h  50 min
  8 h 107 min
```
Add the minutes. Then add the hours.

There are more than 60 minutes.
Rename the minutes as hours and minutes.   107 min = 1 h 47 min

8 h 107 min = 8 h + 1 h 47 min or 9 h 47 min

Ricky exercised for 9 hours 47 minutes the first 3 weeks.

More Examples

A.   3 ft 2 in. →   2 ft 14 in.
      − 1 ft 10 in. → − 1 ft 10 in.
                          1 ft 4 in.

Rename 1 ft as 12 in. and add to 2 in.
Subtract inches; then subtract feet.

B.   2 yd 1 ft   No renaming is
      − 1 yd 1 ft   needed to subtract.
        1 yd 0 ft or 1 yd

C.   3.875 L   No renaming is needed
      + 3.235 L   with like metric units.
        7.110 or 7.11 L

Renaming is not needed with like metric units because parts of the units are expressed as decimals.

TRY These

Complete.

1. 5 yd 2 ft = 4 yd ■ ft
2. 4 T 1,234 lb = 3 T ■ lb
3. 7 d = 6 d ■ h
4. 3 qt 1 pt = 2 qt ■ pt
5. 3 lb 26 oz = 4 lb ■ oz
6. 6 wk 24 d = ■ wk ■ d
7. 5 gal 13 qt = ■ gal ■ qt
8. 2 ft 20 in. = ■ ft ■ in.

Add or subtract.

1. 3 ft 3 in.
   + 4 ft 7 in.

2. 4 lb 15 oz
   − 1 lb 10 oz

3. 6 qt 3 c
   + 2 qt 3 c

4. 3 wk 6 d
   + 2 wk 5 d

5. 3 yd 16 in.
   − 2 yd 30 in.

6. 7 min 15 s
   − 3 min 45 s

7. 3.46 m + 2.56 m

8. 1.2 L + 3 L

9. 4.56 kg + 3.457 kg

★10. 4 yd 1 ft 6 in.
   + 1 yd 2 ft 10 in.

★11. 2 gal 2 qt 2 c
   − 1 gal 3 qt 3 c

PROBLEM Solving

12. Beth uses 3 yards 7 inches of ribbon for one gift and 2 yards 5 inches for another. How much ribbon does she use in all?

13. A dump truck weighs 3 tons when full. Its load of limestone weighs 1,500 pounds. How much does the empty truck weigh?

★14. Juan poured 256 mL of lemon juice into a pitcher. Then he added 1 L of water. How much liquid is in the pitcher?

MIXED Review

15. Peter saves the same three coins each day. If he saves $11.16 during May, which three coins does he save?

Compute.

16. $3\frac{2}{5} + 4\frac{2}{5}$

17. $2\frac{2}{3} + 8\frac{4}{5}$

18. $6\frac{7}{9} - 3\frac{5}{9}$

19. $7 - 3\frac{5}{6}$

20. $5 + 2\frac{3}{7}$

21. $11\frac{2}{7} - 4\frac{2}{3}$

Chapter 6 Review

LANGUAGE and CONCEPTS

Write *true* or *false*. If false, rewrite the statement to make it true.

1. When subtracting mixed numbers, first subtract the whole numbers and then subtract the fractions.

2. After adding or subtracting mixed numbers, rename any improper fractions as mixed numbers in simplest form.

3. To find common denominators, list the factors of both denominators.

4. The LCD stands for the least common denominator.

5. Sometimes to subtract mixed numbers, you need to rename the bottom fraction.

SKILLS and PROBLEM SOLVING

Add or subtract. Write each answer in simplest form. (Sections 6.1–6.2)

6. $\frac{1}{8} + \frac{7}{8}$

7. $\frac{2}{5} - \frac{1}{5}$

8. $\frac{8}{9} - \frac{2}{9}$

9. $\frac{4}{10} + \frac{1}{10}$

10. $\frac{3}{5} + \frac{3}{5}$

11. $\frac{2}{4} + \frac{3}{4}$

12. $\frac{7}{12} + \frac{8}{12}$

13. $\frac{13}{15} + \frac{14}{15}$

Add or subtract. (Sections 6.3, 6.4–6.7, 6.9–6.10)

14. $\frac{11}{12} - \frac{1}{4}$

15. $\frac{6}{7} + \frac{1}{4}$

16. $1\frac{1}{3} + 5\frac{5}{8}$

17. $3\frac{1}{9} + 7\frac{6}{7}$

18. $5\frac{3}{4} - 2\frac{1}{8}$

19. $14\frac{9}{10} - 3\frac{1}{3}$

20. $25\frac{14}{15} - 7\frac{1}{2}$

21. $23\frac{1}{8} - 5\frac{4}{5}$

Compare using <, >, or =. (Section 6.4)

22. $\frac{2}{5} \bullet \frac{3}{7}$

23. $\frac{13}{15} \bullet \frac{8}{9}$

24. $\frac{4}{11} \bullet \frac{5}{9}$

25. $2\frac{2}{3} \bullet \frac{16}{6}$

26. $4\frac{3}{5} \bullet \frac{24}{6}$

27. $1\frac{7}{8} \bullet \frac{17}{9}$

Compute. (Sections 6.11–6.12)

28. $2\frac{1}{2}$ lb = _____ oz

29. 39 ft = _____ yd

30. 3 gal = _____ c

31. $2\frac{3}{4}$ T = _____ lb

32. $4\frac{1}{2}$ mi = _____ yd

33. 50 oz = _____ lb

34. 14 c = _____ pt

35. 45 c = _____ gal

36. 7 wk 30 d = _____ wk _____ d

37. 4 gal 20 qt = _____ gal _____ qt

38. 12 T 500 lb = 11 T _____ lb

39. 4 yd 12 ft = 3 yd _____ ft

40. 5 ft 6 in. + 4 ft 8 in. = _____

41. 3 gal 4 qt – 1 gal 2 qt = _____

42. 4 mi 300 ft + 5 mi 5,000 ft = _____

43. 4 gal 4 qt 5 c + 8 gal 5 qt 6 c = _____

Solve. If there is not enough information, tell what information is needed. If there is too much information, state the extra information. (Section 6.8)

44. Michael is starting a dog-walking business. For every $\frac{1}{2}$-hour walk, he charges $2.50. During this time, he walks most dogs at least $\frac{3}{4}$ of a mile. How many walks does he need to make $35.00?

45. How many times will Michael need to take the dogs around the local park to walk them $1\frac{1}{2}$ miles?

Chapter 6 Test

Add or subtract. Write each answer in simplest form.

1. $13\frac{1}{6} + 8\frac{3}{6}$
2. $9\frac{3}{8} + 3\frac{5}{8}$
3. $11\frac{3}{12} + 12\frac{7}{12}$
4. $\frac{4}{5} + \frac{3}{5}$
5. $\frac{2}{9} + \frac{3}{5}$
6. $20\frac{11}{12} - 14\frac{1}{13}$
7. $\frac{1}{6} + \frac{3}{5}$
8. $6 - 2\frac{3}{7}$
9. $\frac{4}{5} - \frac{1}{8}$
10. $5\frac{1}{2} - 1\frac{3}{4}$

Compare using <, >, or =.

11. $\frac{9}{10} \bullet \frac{12}{14}$
12. $\frac{3}{7} \bullet \frac{2}{5}$
13. $\frac{7}{8} \bullet \frac{8}{9}$
14. $1\frac{1}{3} \bullet \frac{6}{5}$

Compute.

15. $4\frac{1}{4}$ lb = _____ oz
16. 15 ft = _____ yd
17. 2 gal = _____ c
18. $1\frac{1}{2}$ mi = _____ yd
19. 20 oz = _____ lb
20. $3\frac{1}{2}$ lb = _____ oz
21. 3 ft 9 in. + 1 ft 22 in. = _____
22. 5 gal 3 qt − 2 gal 4 qt = _____
23. 2 mi 1,000 ft + 1 mi 6,000 ft = _____
24. 8 gal 3 qt 6c + 1 gal 4 qt 5 c = _____

Solve. If there is not enough information, tell what information is needed. If there is too much information, state the extra information.

25. Tory hiked $1\frac{1}{3}$ miles out to a waterfall. Then, he hiked 72 feet to reach the top of the waterfall. If the hike lasted a total of 3 hours, how far was his round trip if he hiked back the way he came?

26. Tory stopped to rest $\frac{1}{3}$ of the way up the rocks to the waterfall. He rested for $\frac{1}{4}$ hour and then resumed hiking. How long did it take him to reach the top of the waterfall?

Change of Pace

Patterns

What is unusual about this number chart? Are the numbers in each row ordered in the same way?

Find the sums.

1. First column
   $1 + 20 + 21 + 40 + 41 + 60 + 61 + 80 + 81 + 100$

2. First row
   $1 + 2 + 3 + 4 + 5 + 6 + 7 + 8 + 9 + 10$

3. Second row
   $20 + 19 + 18 + 17 + 16 + 15 + 14 + 13 + 12 + 11$

4. Third row
   $21 + 22 + 23 + 24 + 25 + 26 + 27 + 28 + 29 + 30$

1	2	3	4	5	6	7	8	9	10
20	19	18	17	16	15	14	13	12	11
21	22	23	24	25	26	27	28	29	30
40	39	38	37	36	35	34	33	32	31
41	42	43	44	45	46	47	48	49	50
60	59	58	57	56	55	54	53	52	51
61	62	63	64	65	66	67	68	69	70
80	79	78	77	76	75	74	73	72	71
81	82	83	84	85	86	87	88	89	90
100	99	98	97	96	95	94	93	92	91

Solve.

5. Do you think the sum of the numbers in the second column will be greater than, less than, or equal to the sum of the first column?

6. Add the numbers in the second column. Was your guess in problem 5 correct? Can you explain the result?

7. What do you think will be the sum of the numbers in the tenth column?

8. In problems 2–4 you found the sums of numbers in three rows. What will be the sum of the numbers in the fourth row?

9. How can you predict the sum of the numbers in the sixth row? in any row?

10. Do you think the sums of the diagonals (↙ ↗) will be the same? Find them.

11. Look for patterns and special sums of numbers in the chart. Discuss the patterns found.

12. Make an unusual number chart of your own. What patterns does it have?

Cumulative Test

1. $3\frac{3}{5} + 1\frac{8}{15} =$ _____
   a. $4\frac{11}{20}$
   b. $4\frac{2}{15}$
   c. $4\frac{11}{15}$
   d. $5\frac{2}{15}$

2. $16.7 - 1.49 =$ _____
   a. 1.8
   b. 18.19
   c. 15.39
   d. 15.21

3. How many milliliters are in 5 liters?
   a. 50 mL
   b. 500 mL
   c. 5,000 mL
   d. 50,000 mL

4. $\begin{array}{r} 9.037 \\ \times 0.6 \\ \hline \end{array}$
   a. 0.54222
   b. 5.4222
   c. 54.222
   d. none of the above

5. $\begin{array}{r} 2\text{ h }30\text{ min} \\ +\ 3\text{ h }50\text{ min} \\ \hline \end{array}$
   a. 5 h 20 min
   b. 6 h
   c. 6 h 30 min
   d. none of the above

6. $16.05 \div 5 =$ _____
   a. 0.321
   b. 3.021
   c. 32.1
   d. none of the above

7. Janice is using a soup recipe that calls for $1\frac{1}{2}$ cups of heavy cream. She wants to make 4 times the amount the recipe shows. Which of the following measurements is *not* 4 times the amount of cream?
   a. 2 qt
   b. 6 c
   c. 3 pt
   d. 1 qt 1 pt

8. Estimate the length of the eraser in both centimeters and millimeters.

   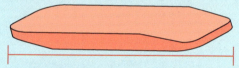

   a. 2 cm, 20 mm
   b. 6 cm, 60 mm
   c. 6 cm, 600 mm
   d. none of the above

9. A large paper clip is about 5 centimeters in length. Josephine made a chain by hooking together 200 large paper clips. About how long was her chain?
   a. 1 m
   b. 10 m
   c. 100 m
   d. 1 km

10. Tom weighs his cat, Sylvester, on a scale that shows ounces. Sylvester weighs 144 ounces. How many pounds does the cat weigh?
    a. 9 lb
    b. $10\frac{1}{2}$ lb
    c. 12 lb
    d. 14 lb

CHAPTER 7

Multiplying and Dividing Fractions

Ashley Gilmore
Davidsonville, MD

7.1 Multiplying Whole Numbers and Fractions

Objective: to multiply fractions and whole numbers

Sandra has 9 pencils in her desk. If $\frac{2}{3}$ of them are red, how many are red?

You can find the answer to this problem by drawing a model or by using multiplication.

Drawing a Model

Multiplying

Step 1	Step 2	Step 3
Write the whole number as a fraction. $\frac{2}{3} \times \frac{9}{1}$	Multiply the numerators. Multiply the denominators. $\frac{2 \times 9}{3 \times 1} = \frac{18}{3}$	Rename the answer. $\frac{18}{3} = 6$

6 pencils are red.

Examples

A. $\frac{1}{5} \times 10 = ?$

$\frac{1}{5} \times \frac{10}{1}$

$\frac{1 \times 10}{5 \times 1} = \frac{10}{5}$

$\frac{10}{5} = 2$

B. $\frac{3}{4} \times 20 = ?$

$\frac{3}{4} \times \frac{20}{1}$

$\frac{3 \times 20}{4 \times 1} = \frac{60}{4}$

$\frac{60}{4} = 15$

TRY These

Multiply. Write each product as a whole number.

1. 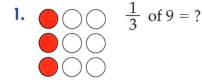 $\frac{1}{3}$ of 9 = ?

2. 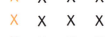 $\frac{1}{4}$ of 16 = ?

3. $\frac{1}{2}$ of 18

4. $\frac{2}{5}$ of 10

5. $\frac{2}{3} \times 15$

Exercises

Copy and complete. Draw pictures if necessary.

1. $\frac{1}{2}$ of 12
2. $\frac{1}{6}$ of 12
3. $\frac{1}{7}$ of 14
4. $\frac{1}{5}$ of 10
5. $\frac{1}{8}$ of 16

6. $\frac{1}{2} \times 22$
7. $\frac{2}{5} \times 15$
8. $\frac{3}{10} \times 30$
9. $\frac{1}{5} \times 25$
10. $\frac{2}{3} \times 9$

11. $\frac{5}{6} \times 12$
12. $\frac{4}{5} \times 10$
13. $\frac{1}{3} \times 21$
14. $\frac{1}{3} \times 12$
15. $\frac{3}{4} \times 28$

Problem Solving

16. When the Smiths went to the market they spent $90 for food. If $\frac{1}{10}$ of that money was spent on snacks, how much money was spent on snacks?

17. In a class of 30 students, $\frac{3}{5}$ are boys. How many are boys?

18. Bob has 24 cards in his collection. If $\frac{1}{3}$ of his cards show Orioles players, how many Orioles cards does he have?

19. Yong said he would bring $\frac{2}{3}$ of the 15 cold wraps for the party and Ethan said he would bring $\frac{1}{2}$ of the 12 hot wraps. Who will bring more wraps?

Constructed Response

20. Out of the 24 students, $\frac{1}{4}$ received a perfect score on the last math test. How many students received a perfect score? Draw a picture and explain.

7.1 Multiplying Whole Numbers and Fractions

7.2 Multiplying Fractions

Objective: to find the product of two fractions

For lunch, Mary ordered Giuseppe's famous rectangular pizza. Mary took $\frac{1}{3}$ of the pizza. Then she decided that was too much, so she gave $\frac{1}{2}$ of her piece to Mark. Use the exploration exercise to find out what part of the pizza Mark received.

Exploration Exercise

Use an area model to explore multiplication of fractions.

1. Draw a rectangle 3 units by 2 units. Label each side with thirds and halves.

2. The rectangle has been divided into 6 equal parts. What fraction of the area of the rectangle does each part represent?

3. Shade $\frac{1}{3}$ of the rectangle with one color and $\frac{1}{2}$ with another color. How many small rectangles are shaded with both colors?

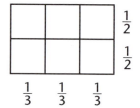

Different Ways to Multiply Fractions

Multiply. $\frac{2}{3} \times \frac{3}{4}$

Use a model.

Step 1	Step 2	Step 3
Draw a 3-unit by 4-unit square and label each side with thirds and fourths.	Shade $\frac{2}{3}$ of the rectangle with one color and $\frac{3}{4}$ with another color.	Write the fraction that is represented by the overlapped area. $\frac{6}{12} = \frac{1}{2}$

192   7.2 Multiplying Fractions

Multiply first; then simplify.

Step 1	Step 2
Multiply the numerators. Multiply the denominators. $\frac{2}{3} \times \frac{3}{4} = \frac{2 \times 3}{3 \times 4} = \frac{6}{12}$	Simplify if necessary. $\frac{6}{12} = \frac{1}{2}$

Simplify first; then multiply.

Step 1	Step 2	Step 3
Look for common factors in the diagonals. $\frac{2}{3} \times \frac{3}{4}$	Cancel any common factors. $\frac{\cancel{2}^{1}}{\cancel{3}_{1}} \times \frac{\cancel{3}^{1}}{\cancel{4}_{2}}$	Multiply. Simplify if necessary. $\frac{1}{1} \times \frac{1}{2} = \frac{1}{2}$

More Examples

A. $\frac{1}{\cancel{4}_{1}} \times \frac{\cancel{8}^{2}}{9} = \frac{2}{9}$

B. $\frac{\cancel{5}^{1}}{6} \times \frac{7}{\cancel{10}_{2}} = \frac{7}{12}$

C. $\frac{2}{\cancel{9}_{3}} \times \frac{\cancel{3}^{1}}{5} = \frac{2}{15}$

TRY These

Multiply. Write each product in simplest form.

1. $\frac{1}{2} \times \frac{1}{6}$

2. $\frac{1}{3} \times \frac{1}{4}$

3. $\frac{2}{5} \times \frac{3}{4}$

4. $\frac{2}{7} \times \frac{3}{4}$

5. $\frac{1}{8} \times \frac{3}{4}$

6. $\frac{1}{3} \times \frac{4}{5}$

Exercises

Multiply. Write each product in simplest form.

1. $\frac{3}{4} \times \frac{1}{6}$
2. $\frac{2}{3} \times \frac{3}{8}$
3. $\frac{4}{9} \times \frac{1}{6}$
4. $\frac{3}{8} \times \frac{3}{4}$
5. $\frac{7}{9} \times \frac{3}{8}$
6. $\frac{2}{5} \times \frac{6}{7}$
7. $\frac{4}{9} \times \frac{1}{2}$
8. $\frac{1}{15} \times \frac{3}{4}$
9. $\frac{2}{3} \times \frac{6}{7}$
10. $\frac{2}{9} \times \frac{5}{6}$
11. $\frac{3}{5} \times \frac{1}{6}$
12. $\frac{3}{14} \times \frac{1}{3}$
13. $\frac{6}{11} \times \frac{5}{6}$
14. $\frac{8}{9} \times \frac{3}{4}$
15. $\frac{1}{8} \times \frac{7}{9}$

PROBLEM Solving

16. Mary budgets her time and allows $\frac{1}{2}$ hour to clean up the kitchen. If $\frac{1}{2}$ of her kitchen cleaning time is used to wash dishes, how much time is spent washing dishes?

17. Tom gave $\frac{1}{3}$ of his baseball cards to his sister Angela. Angela gave $\frac{3}{8}$ of the cards to her friend Sam. What fraction of the cards did Sam receive?

★18. Mrs. Simms baked a pan of brownies. At lunch $\frac{3}{7}$ of the brownies in the pan was eaten and $\frac{1}{4}$ was eaten at dinner. Tina ate $\frac{1}{3}$ of the remaining brownies. How much did Tina eat?

MIND Builder

Logical Reasoning

Complete the mixed-up multiplication chart by using each of the digits 0–9 as factors.

The factor 6 and some of the products are already in place.

×		6		
15			24	
	18		9	
		24		
	0		0	
35			56	

194   7.2 Multiplying Fractions

Cumulative Review

Write each fraction as a decimal.

1. $\frac{3}{5}$
2. $\frac{1}{4}$
3. $\frac{3}{8}$
4. $\frac{16}{20}$

Add or subtract. Write each answer in simplest form.

5. $4\frac{1}{5} + 12\frac{7}{10}$
6. $20\frac{2}{3} + 9\frac{7}{8}$
7. $14\frac{5}{8} - 2\frac{1}{4}$
8. $24\frac{1}{10} - 8\frac{3}{4}$
9. $9\frac{1}{2} + 6\frac{11}{15}$
10. $10 - 5\frac{1}{4}$

Multiply.

11. 26×73
12. 345×98
13. 456×124
14. 5.43×100
15. 0.003×10
16. $18.1 \times 1{,}000$
17. 3.45×1.25
18. 0.09×1.5
19. 8.3×3.4

Divide.

20. $266 \div 7$
21. $40 \div 16$
22. $3{,}078 \div 5$
23. $45.8 \div 8$

Solve.

24. Aaron hiked for $2\frac{1}{2}$ hours. How many minutes did he hike?

25. Jay earned $5 each hour doing chores for his neighbors. He also earned $25 baby-sitting. If he earned $65 altogether, how many hours did he spend doing chores?

26. During the past 20 school days, a cafeteria served spaghetti on 3 days, pizza on 4 days, chicken nuggets on 6 days, and cheese sandwiches on 7 days. If this pattern continues, how many days will the cafeteria serve pizza in the next 80 days?

7.3 Multiplying a Mixed Number by a Fraction and by a Whole Number

Objectives: to find the product of a mixed number and a fraction; to find the product of a mixed number and a whole number

A recipe for crab and spinach quesadillas serves 8 people and uses $2\frac{1}{3}$ cups of spinach dip. Jennifer needs to make enough for 6 people. She decides to make $\frac{3}{4}$ of the recipe. How much spinach dip should she use?

Multiply. $\frac{3}{4} \times 2\frac{1}{3}$

Step 1	Step 2	Step 3
Write the mixed number as an improper fraction. $2\frac{1}{3} = \frac{7}{3}$ $\frac{3}{4} \times 2\frac{1}{3} = \frac{3}{4} \times \frac{7}{3}$	Look for common factors to cancel. $\frac{\cancel{3}^1}{4} \times \frac{7}{\cancel{3}_1} = \frac{7}{4}$	Simplify if possible. Write the fraction as a mixed number if necessary. $\frac{7}{4} = 1\frac{3}{4}$

Jennifer needs $1\frac{3}{4}$ cups of spinach dip.

More Examples

Multiply a mixed number by a fraction.

A. $1\frac{2}{3} \times \frac{2}{5} = \frac{\cancel{5}^1}{3} \times \frac{2}{\cancel{5}_1}$
     $= \frac{2}{3}$ Rewrite the mixed number as an improper fraction.

Multiply a mixed number by a whole number.

B. $3\frac{1}{3} \times 5 = \frac{10}{3} \times \frac{5}{1}$ Rewrite the whole number as a fraction with a denominator of 1.
     $= \frac{50}{3}$
     $= 16\frac{2}{3}$ Rewrite the improper fraction as a mixed number.

TRY These

Multiply. Write each product in simplest form.

1. $\frac{3}{5} \times 3\frac{1}{3}$
2. $4\frac{1}{2} \times \frac{2}{3}$
3. $2\frac{1}{3} \times 6$
4. $\frac{3}{4} \times 3\frac{2}{3}$

Exercises

Multiply. Write each product in simplest form.

1. $2\frac{5}{6} \times \frac{1}{4}$
2. $\frac{1}{4} \times 6\frac{2}{5}$
3. $10 \times 5\frac{1}{5}$
4. $\frac{8}{15} \times 4\frac{4}{5}$
5. $7\frac{2}{4} \times \frac{9}{10}$
6. $12 \times 3\frac{1}{4}$
7. $4\frac{5}{12} \times 8$
8. $\frac{6}{11} \times 3\frac{2}{3}$

Compare using <, >, or =.

9. $4\frac{1}{4} \times \frac{3}{4}$ ● $4\frac{4}{5} \times \frac{1}{6}$
10. $\frac{11}{12} \times 3\frac{3}{5}$ ● $2\frac{3}{4} \times 2$

PROBLEM Solving

11. Bret mows $8\frac{1}{2}$ acres of grass for the golf course each Monday and Tuesday. If he mows $\frac{3}{5}$ of the grass on Monday, how many acres does he have to mow on Tuesday?

12. If 1 cup of fruit punch fills $1\frac{1}{4}$ small glasses, how many small glasses can you fill with 12 cups of punch?

★13. Julie has a whole number that is 3 or less and a mixed number smaller than 4. The product of the two numbers is $10\frac{1}{2}$ and the sum of the two numbers is $6\frac{1}{2}$. What are the two numbers?

Constructed Response

14. TJ had $56. He spent $\frac{5}{8}$ of his money on a pair of shoes and $\frac{1}{2}$ of the remainder on a snack. How much did he spend altogether? Explain your answer.

7.4 Problem-Solving Strategy: Logical Reasoning

Objective: to use logical reasoning to solve problems

Bobby, Wanda, Ken, and May lost their yo-yos. Each yo-yo was a different color (blue, red, purple, and green). Use the clues to match each yo-yo to its owner.

- Wanda and the boy with the green yo-yo live in the same apartment building.
- The purple yo-yo is owned by a girl.
- Ken lives four blocks from Wanda.
- Wanda's yo-yo matches her natural hair color.

1. READ You need to match each yo-yo to its owner. You have clues to help you.

2. PLAN Make a table to show what you know and to help you think logically.
List the names along the side of the table.
List the yo-yo colors along the top of the table.

3. SOLVE
- Start with the easiest clue.
  Wanda's yo-yo matches her natural hair color.
  The only color her hair could be is red!
- Write ✓ where the row for Wanda crosses the column for Red.
- Then decide if that answer can help you solve other clues.

  The purple yo-yo is owned by a girl.
  May is the only other girl.

- Continue until you have solved all the clues.

	Blue	Red	Purple	Green
Bobby		✗		
Wanda	✗	✓	✗	✗
Ken		✗		
May		✗		

4. CHECK Read the clues again. Be sure your table shows only one owner for each yo-yo.

Bobby: green   Wanda: red   Ken: blue   May: purple

TRY These

1. List the order in which three people are standing in line.
   - Fred is not last in line.
   - Marcia is in front of the tallest person in line.
   - Al is standing behind Fred.

2. Ed, Scott, and Jim are cousins. Ed is older than Scott. Jim is younger than Ed. Scott is not the youngest. List the cousins from youngest to oldest.

Solve

1. Bill, Yani, and Sue each have a pet. The pets are a dog, a canary, and a hamster. Which pet does each child have?
   - Sue's pet lives in a wire cage.
   - Bill's pet has a long leash.
   - Yani's pet does not have wings.

2. On a picnic, Travis, Hakim, and Niguel each bring a sandwich (ham, pb & j, and turkey) and a drink (water, chocolate milk, and a sports drink) Both Hakim and Travis bring meat. Niguel does not eat chocolate. Travis does not eat pork and cannot have any sugary drinks. The person who brings turkey does not bring water. What does each person bring for lunch?

MID-CHAPTER Review

Multiply. Write each answer in simplest form.

1. $\frac{2}{3} \times 12$
2. $90 \times \frac{2}{3}$
3. $\frac{1}{3} \times \frac{2}{9}$
4. $\frac{2}{5} \times \frac{6}{7}$
5. $\frac{2}{8} \times \frac{2}{9}$
6. $5 \times 2\frac{2}{5}$
7. $40 \times 1\frac{1}{4}$
8. $2\frac{1}{8} \times 3$

7.4 Problem-Solving Strategy: Logical Reasoning

7.5 Multiplying Mixed Numbers

Objective: to find the product of two mixed numbers

Makenzie uses $1\frac{1}{2}$ cups of sugar for each batch of chocolate chip cookies she makes. How much sugar does she need to make $5\frac{1}{3}$ batches of chocolate chip cookies?

Multiply. $1\frac{1}{2} \times 5\frac{1}{3}$

Step 1	Step 2	Step 3
Write each mixed number as an improper fraction. $1\frac{1}{2} = \frac{3}{2}$ $5\frac{1}{3} = \frac{16}{3}$ $1\frac{1}{2} \times 5\frac{1}{3} = \frac{3}{2} \times \frac{16}{3}$	Look for common factors to cancel. $\frac{\cancel{3}^1}{\cancel{2}_1} \times \frac{\cancel{16}^8}{\cancel{3}_1} = \frac{8}{1}$	Simplify if possible. Write the fraction as a mixed number if necessary. $\frac{8}{1} = 8$

Makenzie will need 8 cups of sugar.

Other Examples

A. $2\frac{2}{3} \times 2\frac{1}{4} = \frac{\cancel{8}^2}{\cancel{3}_1} \times \frac{\cancel{9}^3}{\cancel{4}_1}$
$= \frac{6}{1}$
$= 6$

B. $4\frac{1}{8} \times 3\frac{3}{11} = \frac{\cancel{33}^3}{\cancel{8}_2} \times \frac{\cancel{36}^9}{\cancel{11}_1}$
$= \frac{27}{2}$
$= 13\frac{1}{2}$

TRY These

Multiply. Write each product in simplest form.

1. $2\frac{1}{6} \times 2\frac{2}{7}$
2. $1\frac{2}{3} \times 1\frac{3}{4}$
3. $5\frac{1}{4} \times 4$
4. $3\frac{2}{3} \times 3\frac{3}{7}$

Exercises

Multiply. Write each product in simplest form.

1. $1\frac{2}{5} \times 2\frac{1}{3}$
2. $3\frac{2}{3} \times 1\frac{3}{7}$
3. $6\frac{3}{8} \times 3\frac{1}{5}$
4. $6\frac{1}{5} \times 3\frac{1}{4}$
5. $1\frac{5}{6} \times 3\frac{1}{2}$
6. $3\frac{2}{7} \times 1\frac{1}{8}$
7. $4\frac{4}{5} \times 2\frac{3}{4}$
8. $10 \times 1\frac{1}{4}$

Problem Solving

9. Mike is power-washing his deck that is $5\frac{1}{2}$ times as wide as he is tall. If Mike is $5\frac{5}{6}$ feet tall, how many feet wide is Mike's deck?

10. Ciara worked $7\frac{1}{2}$ hours on Friday. Ciara's assistant Clarissa worked $1\frac{1}{5}$ as long as Ciara did. How many hours did Clarissa work on Friday?

11. Ivan has $3\frac{1}{2}$ pages of work to make up because he was absent from school. Ivar was absent longer and has $1\frac{1}{2}$ times as many pages of make up work. How many pages does Ivar need to complete?

12. Can the product of two mixed numbers ever equal a whole number? If so, give an example.

Constructed Response

13. Rachel bought $3\frac{2}{3}$ pounds of peaches for herself from the farmer's market. She bought $1\frac{1}{2}$ times as many pounds for her neighbors. How many pounds did she buy for her neighbors? Explain.

7.6 Dividing a Whole Number by a Fraction

Objective: to use a reciprocal to divide a whole number by a fraction

Kate is making favors for her friend Jenny's birthday party. She has 18 cups of chocolate candies. She wants to put the candies in bags so that there is $\frac{1}{2}$ cup in each bag. How many bags can Kate fill? Think, how many halves are in 18? You solve this problem by dividing: $18 \div \frac{1}{2}$.

To help Kate figure out this problem, use the exercise below.

Exploration Exercise
Use the model to explore division of a whole number by a fraction.

1. Draw 18 wholes.
2. Mark them in halves.
3. Color each $\frac{1}{2}$. Count the halves.

There are 36 halves in 18. But how can you divide any number by a fraction?

To divide a whole number by a fraction, you need to multiply by the **reciprocal**. Reciprocals are number pairs that have a product of 1. To find the reciprocal of a fraction, reverse the numerator and the denominator.

$$\frac{4}{5} \times \frac{5}{4} = 1 \qquad\qquad 4 \times \frac{1}{4} = 1 \qquad\qquad \frac{6}{11} \times \frac{11}{6} = 1$$

Examples

A. $3 \div \frac{5}{6} = 3 \times \frac{6}{5} = \frac{3}{1} \times \frac{6}{5} = \frac{18}{5} = 3\frac{3}{5}$ $\frac{5}{6}$ and $\frac{6}{5}$ are reciprocals.

B. $12 \div \frac{3}{4} = 12 \times \frac{4}{3} = \frac{12}{1} \times \frac{4}{3} = \frac{16}{1} = 16$ $\frac{3}{4}$ and $\frac{4}{3}$ are reciprocals.

C. $8 \div \frac{2}{7} = 8 \times \frac{7}{2} = \frac{8}{1} \times \frac{7}{2} = \frac{28}{1} = 28$ $\frac{2}{7}$ and $\frac{7}{2}$ are reciprocals.

TRY These

Find the reciprocal of each number.

1. $\frac{1}{2}$
2. $\frac{4}{9}$
3. $\frac{3}{2}$
4. 6
5. 10
6. $\frac{8}{4}$

Complete each equation by writing the reciprocal of the divisor.

7. $4 \div \frac{1}{3} = 4 \times \frac{\blacksquare}{\blacksquare}$
8. $7 \div \frac{14}{15} = 7 \times \frac{\blacksquare}{\blacksquare}$
9. $9 \div \frac{18}{5} = 9 \times \frac{\blacksquare}{\blacksquare}$

Exercises

Divide. Write each quotient in simplest form.

1. $15 \div \frac{1}{3}$
2. $5 \div \frac{5}{9}$
3. $4 \div \frac{2}{3}$
4. $10 \div \frac{3}{8}$
5. $12 \div \frac{7}{6}$
6. $4 \div \frac{1}{8}$
7. $16 \div \frac{5}{8}$
8. $20 \div \frac{5}{11}$

PROBLEM Solving

9. Meg had 6 yards of ribbon to make hair ties. If she uses $\frac{1}{5}$ of a yard for each hair tie, how many hair ties can she make?

10. Shelia poured 4 gallons of water equally into some identical bottles. The bottles were filled to the brim and each bottle contained $\frac{1}{10}$ gallon of water. How many bottles did she use?

11. George has $15. If he exchanges the $15 for nickels, how many nickels will he have? (Hint: A nickel is $\frac{1}{20}$ of a dollar.)

★12. Of the students in fifth grade $\frac{1}{5}$ play basketball. If 25 students play basketball, how many students are in the fifth grade?

7.7 Dividing Fractions

Objectives: to divide a fraction by a whole number; to divide a fraction by a fraction

At lunch, Abby wants to share $\frac{3}{4}$ of her bag of carrots with Blaire. You can divide fractions to help you figure out how to share equally.

To solve the problem you can divide $\frac{3}{4}$ by 2.

Using a Model

Step 1	Step 2
Show $\frac{3}{4}$.	Show $\frac{3}{4}$ in two equal parts. Each small rectangle is $\frac{1}{8}$ of the whole. Each of two equal parts of the $\frac{3}{4}$ has three of these eighths.

Dividing

Step 1	Step 2	Step 3
Write the whole number as a fraction. $\frac{3}{4} \div 2 = \frac{3}{4} \div \frac{2}{1}$	Multiply by the reciprocal of the divisor. $\frac{3}{4} \div \frac{2}{1} = \frac{3}{4} \times \frac{1}{2} = \frac{3}{8}$	Write the product in simplest form. $\frac{3}{8}$ is already in simplest form.

Abby can share $\frac{3}{8}$ of her bag of carrots with Blaire.

Examples

A. $\frac{3}{10} \div 5 = \frac{3}{10} \times \frac{1}{5} = \frac{3}{50}$ 5 and $\frac{1}{5}$ are reciprocals.

B. $\frac{4}{5} \div \frac{4}{15} = \frac{\cancel{4}^1}{5} \times \frac{\cancel{15}^3}{\cancel{4}_1} = \frac{3}{1} = 3$ $\frac{4}{15}$ and $\frac{15}{4}$ are reciprocals.

C. $6 \div 9 = \frac{\cancel{6}^2}{1} \times \frac{1}{\cancel{9}_3} = \frac{2}{3}$ 9 and $\frac{1}{9}$ are reciprocals.

TRY These

Complete each equation by writing the reciprocal of the divisor.

1. $\frac{3}{5} \div \frac{4}{9} = \frac{3}{5} \times \frac{\blacksquare}{\blacksquare}$

2. $\frac{5}{9} \div 3 = \frac{5}{9} \times \frac{\blacksquare}{\blacksquare}$

3. $\frac{3}{8} \div \frac{11}{15} = \frac{3}{8} \times \frac{\blacksquare}{\blacksquare}$

Exercises

Divide.

1. $\frac{2}{3} \div 15$

2. $\frac{4}{7} \div \frac{3}{8}$

3. $\frac{2}{5} \div 15$

4. $\frac{3}{8} \div \frac{4}{9}$

5. $\frac{3}{4} \div \frac{1}{5}$

6. $\frac{9}{11} \div \frac{21}{22}$

7. $4\frac{1}{2} \div \frac{9}{10}$

8. $\frac{11}{12} \div \frac{3}{4}$

9. $40 \div \frac{1}{4}$

10. $\frac{2}{7} \div \frac{3}{5}$

11. $\frac{3}{5} \div \frac{5}{6}$

12. $\frac{3}{10} \div \frac{21}{100}$

PROBLEM Solving

13. Mr. Kraft owns $\frac{4}{5}$ of an acre of property. If he splits his property into 4 equal sections to give to his children, how many acres will each child receive?

14. Lee has $1\frac{3}{4}$ pounds of flour. He packs all the flour equally into 12 bags. How much flour is in each bag?

15. Alana has $\frac{3}{4}$ of a yard of fabric. She wants to make change purses for her friends. If each change purse uses $\frac{3}{8}$ of a yard of fabric, how many change purses can Alana make?

Constructed Response

16. Johnny says that $\frac{3}{7} \div \frac{5}{9} = \frac{15}{63}$. Is he correct? Explain.

7.7 Dividing Fractions

7.8 Problem-Solving Strategy: Solve a Simpler Problem

The quilt has 4 colors and 3 different shapes. How many different pieces are there?

Think about solving a simpler problem.

2 colors
1 shape
2 (2 × 1) different pieces

2 colors
2 shapes
4 (2 × 2) different pieces

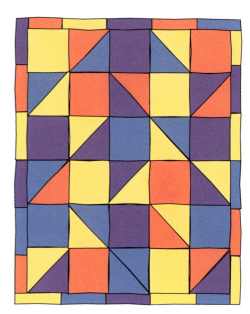

1. READ You know the number of colors and the number of shapes. You need to find the number of different pieces.

2. PLAN From the simpler problems, you can see that multiplying the number of shapes by the number of colors tells the number of different pieces.
Write an equation. $n = 4 \times 3$

3. SOLVE $4 \times 3 = 12$
There are 12 different pieces.

4. CHECK A tree diagram can be used to show all the combinations of shape and color.

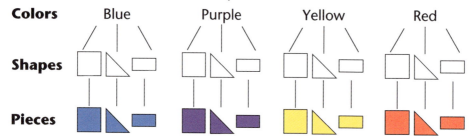

There are 12 different pieces. The answer checks.

TRY These

1. A quilt is made of rectangles, squares, and triangles. The shapes are red, white, and blue. How many different kinds of pieces are there?

2. There were 36 students in a chess tournament. In each round, one student played one match against another student and the winner continued until there was only one student remaining. How many total matches were played?

Solve

1. How many rectangles are shown below?

2. Find the sum of the first 50 odd numbers.

3. At a dinner party, there are 6 guests. Each guest is to share a small gift with each other guest. How many gifts will be exchanged?

4. At swimming practice, you do 1 lap the first day and double the number you do each day for 10 days. How many laps will you do on the tenth day?

5.
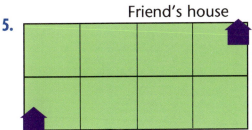

How many blocks do you have to walk from home to your friend's house? How many different ways can you walk this number of blocks?

6. What day of the week is 100 days after Tuesday?

★7. A stack of quarters is worth $100. How many quarters are in the stack?

7.8 Problem-Solving Strategy: Solve a Simpler Problem   **207**

Chapter 7 Review

LANGUAGE and CONCEPTS

Match.

1. fractions in which the numerator is greater than or equal to the denominator
2. two fractions whose product is 1
3. sums of whole numbers and fractions

a. reciprocals
b. mixed numbers
c. improper fractions

SKILLS and PROBLEM SOLVING

Multiply. Write each product in simplest form. (Sections 7.1–7.3, 7.5)

4. $\frac{1}{3} \times \frac{3}{5}$
5. $\frac{1}{3} \times \frac{5}{9}$
6. $\frac{4}{9} \times \frac{12}{28}$
7. $\frac{7}{8} \times \frac{24}{21}$
8. $4\frac{2}{5} \times \frac{5}{11}$
9. $4 \times \frac{12}{28}$
10. $3\frac{1}{2} \times 9$
11. $9\frac{4}{15} \times \frac{3}{8}$
12. $10\frac{3}{4} \times 1\frac{5}{6}$
13. $3\frac{2}{9} \times 8$
14. $2\frac{1}{4} \times 3\frac{1}{3}$
15. $4\frac{1}{5} \times \frac{19}{20}$

Write the reciprocal of each number. (Section 7.6)

16. $\frac{1}{7}$
17. $\frac{5}{8}$
18. $\frac{9}{4}$
19. 8
20. 11
21. $\frac{12}{3}$

208   Chapter 7 Review

Divide. Write each quotient in simplest form. (Sections 7.6–7.7)

22. $\dfrac{1}{3} \div \dfrac{5}{6}$
23. $\dfrac{4}{5} \div \dfrac{9}{10}$
24. $\dfrac{3}{8} \div \dfrac{21}{24}$

25. $\dfrac{2}{9} \div \dfrac{8}{6}$
26. $8 \div \dfrac{6}{7}$
27. $10 \div \dfrac{2}{11}$

28. $15 \div \dfrac{7}{8}$
29. $\dfrac{1}{6} \div 7$
30. $\dfrac{5}{8} \div 20$

31. $\dfrac{3}{4} \div 10$
32. $\dfrac{1}{2} \div 3$
33. $\dfrac{3}{4} \div \dfrac{3}{4}$

Solve. (Sections 7.3, 7.7)

34. Sue works 3 shifts every week. Each shift lasts $4\dfrac{1}{4}$ hours. How many hours does she work in 1 week?

35. Ryan has $3\dfrac{3}{4}$ rolls of tape and needs $\dfrac{2}{3}$ of a roll to tape each tennis racket. How many tennis rackets can he tape?

36. It takes $3\dfrac{3}{4}$ cups of pancake mix to make 12 pancakes. How many cups of pancake mix would you need to make 1 pancake?

37. Without calculating, decide which is greater $\dfrac{3}{4} \times \dfrac{6}{7}$ or $\dfrac{3}{4} \div \dfrac{6}{7}$. Explain how you know.

Chapter 7 Review

Chapter 7 Test

Multiply. Write each product in simplest form.

1. $\dfrac{4}{9} \times \dfrac{1}{2}$
2. $\dfrac{1}{4} \times \dfrac{3}{8}$
3. $\dfrac{2}{5} \times \dfrac{3}{4}$
4. $4 \times \dfrac{5}{16}$
5. $\dfrac{3}{10} \times 70$
6. $\dfrac{4}{9} \times \dfrac{6}{16}$
7. $\dfrac{5}{6} \times 2\dfrac{1}{8}$
8. $6\dfrac{3}{4} \times \dfrac{8}{9}$
9. $5\dfrac{9}{10} \times 1\dfrac{1}{9}$

Write the reciprocal of each number.

10. $\dfrac{1}{4}$
11. $\dfrac{2}{9}$
12. $\dfrac{14}{3}$
13. 10

Divide. Write each product in simplest form.

14. $\dfrac{3}{4} \div 3$
15. $\dfrac{8}{9} \div 5$
16. $\dfrac{3}{5} \div 11$
17. $2 \div \dfrac{4}{5}$
18. $7 \div \dfrac{21}{22}$
19. $16 \div \dfrac{1}{4}$
20. $\dfrac{2}{5} \div \dfrac{3}{25}$
21. $\dfrac{24}{25} \div \dfrac{6}{5}$
22. $\dfrac{7}{9} \div \dfrac{63}{12}$
23. $\dfrac{15}{18} \div \dfrac{3}{9}$
24. $\dfrac{1}{8} \div \dfrac{5}{12}$
25. $3 \div 18$

Solve.

26. Find *three* pairs of fractions that each have a product of $\dfrac{1}{2}$.

27. A roll of cookie dough with a length of $\dfrac{3}{4}$ feet is cut into 24 equal pieces. What is the length of each piece in feet?

28. Mr. Holt used $\dfrac{2}{5}$ of a pound of sugar for his coffee in 8 days. If he used the same amount each day, how much sugar did he use each day? Give your answer in pounds.

29. A rectangle measures $\dfrac{2}{6}$ meter by $\dfrac{5}{7}$ meter. Find its area. (*Remember:* To find area multiply length times width.)

Change of Pace

Optical Illusions

Our eyes can sometimes fool us. Certain arrangements of points, lines, and planes create optical illusions because our mental image is different from the real image. In geometry you must learn to question your own mental images.

Look at this figure. Which segment seems to be longer, the horizontal one (from C to D) or the vertical one (from A to B)?

Use a ruler to measure the two segments. What did you discover?

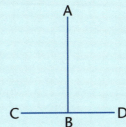

First use your mental image to guess an answer to each question. Then, use a ruler or a straightedge to see if you were right.

1. Are the sides of the triangle straight or curved?

2. Which of the three segments at the top forms a straight line with the segment at the bottom?

   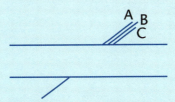

3. Which segment is the longest?

4. Which is longer, the segment from A to B or the segment from B to D?

   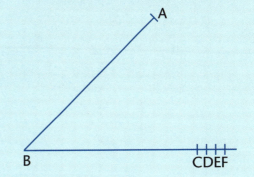

Change of Pace 211

Cumulative Test

1. $\frac{3}{4} + \frac{2}{3} =$ _____
   a. $\frac{1}{2}$
   b. $\frac{5}{7}$
   c. $1\frac{5}{12}$
   d. $1\frac{1}{2}$

2. How is the improper fraction $\frac{18}{12}$ written in simplest form?
   a. $\frac{2}{3}$
   b. $\frac{9}{6}$
   c. $1\frac{1}{3}$
   d. $1\frac{1}{2}$

3. Which set of fractions is in order from least to greatest?
   a. $\frac{1}{2}$ $\frac{1}{4}$ $\frac{1}{8}$
   b. $\frac{2}{3}$ $\frac{3}{4}$ $\frac{5}{6}$
   c. $\frac{1}{5}$ $\frac{2}{5}$ $\frac{1}{4}$
   d. $\frac{4}{6}$ $\frac{7}{8}$ $\frac{10}{12}$

4. 3.481 + 48 = _____
   a. 3.529
   b. 51.481
   c. 8.281
   d. 3.961

5. 20.030 − 6.5 = _____
   a. 19.965
   b. 13.53
   c. 26.530
   d. 20.095

6. $4\frac{7}{8} - 1\frac{3}{4} =$ _____
   a. $2\frac{1}{8}$
   b. $3\frac{3}{4}$
   c. $3\frac{7}{8}$
   d. $3\frac{1}{8}$

7. A restaurant has 42 tables. Each table seats 4 people. If every table is full, how many people are seated in the restaurant?
   a. 46
   b. 38
   c. 168
   d. 172

8. Ms. Beatty read $\frac{2}{9}$ of a book yesterday and $\frac{1}{3}$ of the book today. How much of the book has she read in all?
   a. $\frac{3}{12}$ of the book
   b. $\frac{5}{9}$ of the book
   c. $\frac{9}{18}$ of the book
   d. all of the book

9. Alison frosts a cake using $\frac{2}{3}$ cup of frosting. She cuts the cake into 4 equal pieces and eats 1 of them. How much frosting does she eat?
   a. $\frac{2}{3}$ cup
   b. $\frac{1}{2}$ cup
   c. $\frac{3}{8}$ cup
   d. $\frac{1}{6}$ cup

10. Which relationship between units is correct?
    a. One ounce is $\frac{1}{16}$ of one pound.
    b. One pound is $\frac{1}{16}$ of one ounce.
    c. One cup is of $\frac{1}{8}$ one gallon.
    d. One gallon is $\frac{1}{8}$ of one cup.

CHAPTER 8

Geometry

James Miller
Calvert Day School

8.1 Geometry Basics

Objectives: to identify and name points, lines, line segments, rays, and angles; to use parallel lines

There are many geometric figures in everyday items. You can find points, lines, planes, and angles in items such as roads, tabletops, pencils, and sunbeams. Many important terms in geometry are listed in the tables below.

Geometry Basics			
Line	A straight path of two points that continues on forever in two directions. It is named by two points.		Line AB or BA $\overleftrightarrow{AB}$ or $\overleftrightarrow{BA}$
Line Segment	A part of a line. It is made up of two endpoints and all of the points on the line that are between them.		Line segment CD or DC $\overline{CD}$ or $\overline{DC}$
Ray	A ray is formed by an endpoint and all of the points that continue forever in a certain direction from that endpoint.		Ray MN $\overrightarrow{MN}$
Angle	Two rays that have the same endpoint form an angle.		Angle BAC ∠BAC

Parallel and Perpendicular Lines			
Intersecting Lines	Two lines that cross once.		Line EF intersects line JK.
Parallel Lines	Two lines that are in the same plane, but never cross.		Line AB is parallel to line CD. AB∥CD
Perpendicular Lines	Lines that intersect to form a right angle.		Line EF is perpendicular to line GH. EF⊥GH

TRY These

Tell whether each situation describes a *line*, a *line segment*, a *ray*, or an *angle*.

1. the hour and minute hands on a clock
2. a beam of light
3. light from the Sun
4. a pencil

Exercises

Use the figure to the right to answer questions 1–4.

1. How many rays do you see? List them.
2. Find four angles. Name them.
3. Do $\overrightarrow{CF}$ and $\overrightarrow{CD}$ form an angle?
4. Name the rays that form ∠DCG.

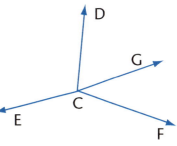

Use the figure at the right to answer questions 5–7.

5. Name four angles formed by the perpendicular lines.
6. Find four rays. Name them.
7. Find examples of parallel lines in your classroom.

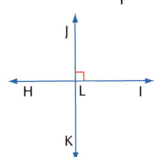

Problem Solving

8. Julia is driving south on Charles Street. Nicole is driving north on Charles Street. Their paths suggest _____.

   a. parallel lines   b. perpendicular lines

9. Mario begins to cut his pizza by starting on one edge and slicing through the center to the opposite side. He then turns the pizza a quarter turn and cuts again, in the same manner. His first two cuts suggest _____.

   a. parallel lines   b. perpendicular lines

Mixed Review

Compute.

10. $3.92 × 47
11. 22 × 2,603
12. 5,370 × 80
13. 9)3,468
14. 4)8,621
15. 505 ÷ 65

8.1 Geometry Basics 215

8.2 Angles

Objectives: to estimate angle measures; to name and classify angles; to measure and draw angles

Surveying is used to locate property lines and routes of highways. Namid Hawkes uses a transit to measure angles. The transit turns in a complete circle.

Units called **degrees** (°) are used to measure angles. A full turn of the transit is a 360° turn. A minute hand on a clock makes a 360° turn every hour.

Suppose ray AB turns about the circle. Think of ray AB stopping at positions that divide the circle into four equal parts. Each part forms a turn of 90°.

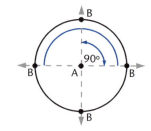

Each new position of ray AB forms an angle with its starting position. The first stop forms a 90° angle. The second stop is halfway around the circle at 180°.

Angles of 90° are called **right angles**. Other angles can be compared to 90° angles. Right angles can be shown with the symbol ⌐.

Examples

Tell whether each measure is a good estimate. Why or why not?

A. 45°

B. 175°

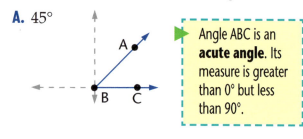

Angle ABC is an **acute angle**. Its measure is greater than 0° but less than 90°.

Angle DEF is an **obtuse angle**. Its measure is greater than 90° but less than 180°.

∠ABC is about half a right angle. 90 ÷ 2 = 45
45° is a good estimate.

∠DEF is a little greater than a right angle (90°).
175° is not a good estimate.
A better estimate is 100°.

Architects often measure and draw angles when they provide detailed plans for builders. A **protractor** can be used to measure and draw angles.

More Examples

C. Measure ∠ABC.

- Place the center mark of the protractor on the vertex (B).
- Place the 0° mark of the protractor on one side of the angle (ray BC).
- Read the measure of the angle where the other side crosses the scale.

The measure of ∠ABC is 70°.

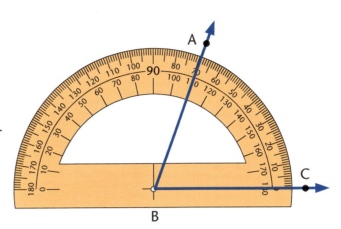

The scale shows 70° and 110°. Since ∠ABC is less than a right angle, use 70°.

D. Draw ∠PQR whose measure is 110°.

- Draw one side of the angle (ray QP).
- Place the protractor on the side so that the vertex (Q) is at the center point and the side (ray QP) goes through the 0° mark.
- Look along the scale that starts with 0°. Find 110° and place a point at the edge of the protractor.
- Remove the protractor. Draw the other side of the angle by drawing a ray (QR) from the vertex through the 110° point.

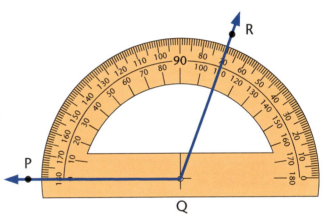

TRY These

Name the kind of angle—*acute*, *right*, or *obtuse*. Then write each angle measurement.

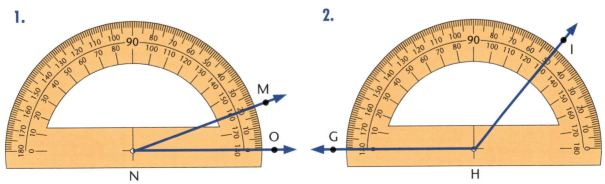

8.2 Angles **217**

Exercises

Choose the correct angle measurement.

1.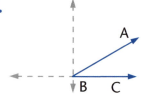
a. 80°   b. 90°   c. 30°

2.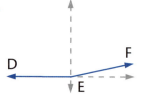
a. 90°   b. 170°   c. 70°

3.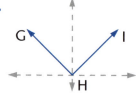
a. 45°   b. 150°   c. 90°

Estimate. Then measure each angle.

4.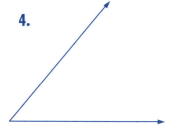

5.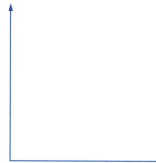

6.

Draw an angle for each given measure.

7. 50°   8. 140°   9. 25°   10. 105°

PROBLEM Solving

11. Each day the sun rises in the east and sets in the west. About how many degrees does it seem to travel?

12. If a star moves 1° across the sky every 4 minutes, how many degrees does it move in 1 hour? (1 h = 60 min)

13. The hour and minute hand of a clock at 3:00 suggest what kind of an angle?

14. Suppose a pizza is cut across the middle to form 8 equal pieces. What is the measure of the angle of each piece?

15. How many angles are in the figure?

Cumulative Review

Write in standard form and expanded form.

1. three thousand, six hundred seventy-four
2. one million, eight hundred thousand

Write the value of the underlined digit.

3. 2_7_6
4. _7_,360
5. 1,56_7_
6. 8,_7_64

Compare using <, >, or =.

7. 36,000 ● 3,600
8. 2,791 ● 2,801
9. 89,352 ● 89,352

Compute.

10. 2,603 + 395
11. 9,607 + 2,394
12. 411 − 199
13. $150 − 37
14. 7,053 − 6,317

15. 71 × 3
16. $2.40 × 5
17. 38 × 41
18. 269 × 24
19. 583 × 406

20. 4)80
21. 5)265
22. 8)9,621
23. 20)1,025
24. 31)1,426

Estimate.

25. 793 + 826
26. 282 − 198
27. 919 × 63
28. 481 × 312
29. 6)312

Solve.

30. On Wednesday 6 cars were driven to the fair carrying 4 people each. The next day 4 cars were driven to the fair carrying 6 people each. What property does this show?

31. Mr. Jeffries is a landscaper. He plants 14 shrubs on each house lot. How many shrubs does he plant on 32 lots?

32. Jeff earns $8.00 doing yard work and $3.75 baby-sitting. How much more must he earn to have $15.00?

33. Rico and 2 friends picked 69 pints of cherries. If each person took home the same amount, how many pints did Rico take home?

8.3 Triangles

Objectives: to classify triangles; to find missing angle measures

There are a number of different types of triangles. You can classify triangles by their angles.

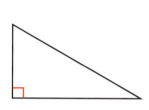

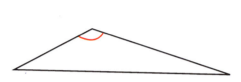

A **right triangle** has one 90° angle.

An **acute triangle** has all three angles less than 90°.

An **obtuse triangle** has one angle that is greater than 90°.

You can also classify triangles by their sides.

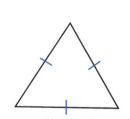

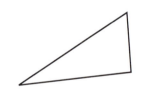

An **equilateral triangle** has three sides that are the same length.

An **isosceles triangle** has two sides that are the same length.

A **scalene triangle** has no side that is the same length as any other.

Examples

A. Classify the triangle in two ways.

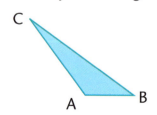

Triangle ABC has one angle that is obtuse.
Triangle ABC has no side that is the same length as any other side.

Triangle ABC is an obtuse scalene triangle.

220   8.3 Triangles

B. Classify the triangle in two ways.

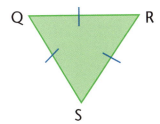

Triangle QRS has all acute angles.
Triangle QRS has all sides that are the same length.

Triangle QRS is an acute equilateral triangle.

Jack knows there is a special rule for the measures of the angles of a triangle. This rule works for triangles of any shape or size.

Use a ruler to draw any triangle. Cut it out. Write the numbers 1, 2, and 3 to number each angle.

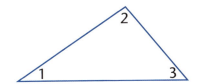

Tear the angles off the triangle. Arrange the pieces as shown below. What degree measure do they show?

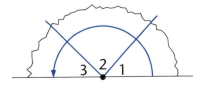

The sum of the pieces shows 180°.

You can also measure each angle of a triangle with a protractor. Then find the sum of the measures.

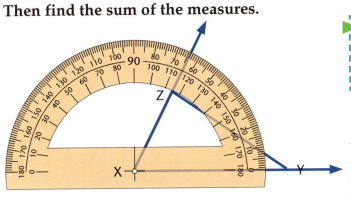

▶ The sides of the triangle are parts of rays that form the angles. Use a ruler to draw the rays so you can measure each angle.

∠X measures 65°.
∠Y measures 40°.
∠Z measures 75°.

65 + 40 + 75 = 180

The sum of the measures is 180°.

Triangle XYZ has all acute angles. Each angle will have a measure less than 90°.

The sum of the measures of the angles of any triangle is 180°.

8.3 Triangles 221

TRY These

Classify each triangle by its sides and angles.

1.
2.
3.
4.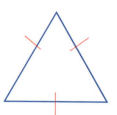

Exercises

Trace each triangle. Then measure the angles in each triangle.

1.
2.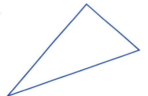
3.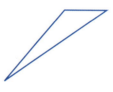

The measures of two angles of a triangle are given. Find the measure of the third angle.

4. 35°, 85°
5. 90°, 40°
6. 120°, 45°
7. 60°, 60°
8. 45°, 45°
9. 70°, 60°

PROBLEM Solving

10. A triangle has a right angle. One acute angle has a measure of 60°. What is the measure of the other angle?

Constructed Response

11. A triangle has three angles that have equal measures. What is the measure of each angle? Explain how you solved this.

MIXED Review

Solve.

12. Paul has $188 in his savings account. If he saves $6 each week, how long will it take for him to have $500 in his account?

13. There are 154 pairs of children's shoes on the shelves. Each shelf holds 11 pairs. How many shelves of children's shoes are there?

14. The first three triangular numbers are shown at right. Name the next *two* triangular numbers.

Replace each ■ to make a true equation. Write the name of the property that you used.

15. 23 × 47 = ■ × 23

16. (2 × 13) × 56 = 2 × (13 × ■)

17. 3,456 × ■ = 3,456

18. ■ × 55 = 0

19. 3 × 9 = (3 × 6) + (3 × ■)

20. 15 × 17 = 17 × ■

8.4 Polygons

Objectives: to determine whether a figure is a polygon; to identify polygons by the number of sides

Jake Carlyle designs and installs tile patios. The tiles he uses are models of **polygons**.

A polygon is a closed figure in a plane. It is made up of line segments that meet but do not cross.

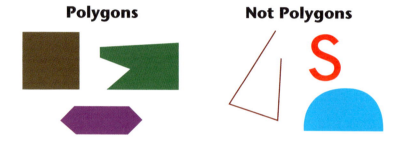

Polygons are named by their number of sides or angles as shown in the chart to the right.

Polygon	Angles	Sides	Figure
Triangle	3	3	△
Quadrilateral	4	4	▭
Pentagon	5	5	⬠
Hexagon	6	6	⬡
Octagon	8	8	⯃

If all the sides and angles of a polygon are congruent (the same size), the polygon is called a **regular polygon**.

A **square** is an example of a regular polygon because it has four equal sides and four equal angles.

TRY These

Tell whether or not each figure below is a polygon. Write *yes* or *no*.

1.
2.
3.
4.
5.
6.

Exercises

Classify each polygon as a *triangle, quadrilateral, pentagon, hexagon,* or *octagon*.

1.
2.
3.
4.

PROBLEM Solving

5. Identify all polygons in the figure below.

6. What do the quadrilaterals shown below have in common?

7. Draw *four* different pentagons.

8. Study the four pentagons you made in problem 7. Describe what you did to make the figures different.

Constructed Response

9. Do any of your pentagons from problem 7 have sides that are the same length? Explain how you know they have the same length.

8.4 Polygons   225

8.5 Quadrilaterals

Objective: to name and classify quadrilaterals

When playing basketball, when Olivia shoots from the free-throw line, she is standing at one side of a **quadrilateral**.

The shape of the free-throw lane is called a rectangle. Trapezoids, parallelograms, rhombi, and squares are all examples of quadrilaterals. A quadrilateral is a polygon with four sides and four angles.

Name	Description	Example	
Trapezoid	exactly one pair of sides is parallel		1 pair of parallel sides
Parallelogram	opposite sides are parallel and equal in length		2 pairs of equal sides 2 pairs of parallel sides
Rectangle	opposite sides are parallel and has four right angles		2 pairs of parallel sides 4 right angles
Rhombus	opposite sides are parallel and all sides are equal in length		4 equal sides 2 pairs of parallel sides
Square	rectangle with all sides the same length		4 equal sides 4 right angles

Many quadrilaterals can be described using more than one name. The Venn diagram below shows the relationships among different quadrilaterals.

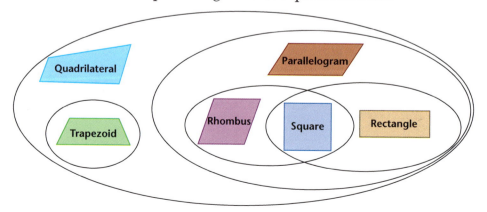

TRY These

1. Which quadrilaterals have four right angles?
2. Which quadrilaterals have all sides equal?
3. How are a square and a rectangle alike? How are they different?

Exercises

Draw and name the quadrilateral.

1. one pair of parallel sides
2. four sides equal and four right angles
3. four sides of any length
4. opposite sides are parallel; two pairs of equal sides

Write *yes* or *no*.

5. A rhombus has four equal sides.
6. A quadrilateral can have sides of any length.
7. A trapezoid has four equal sides.
8. A parallelogram has two pairs of parallel sides.
9. A square has two acute and two obtuse angles.
10. All four-sided polygons are quadrilaterals.

PROBLEM Solving

11. The distance between bases on a rhombus-shaped softball field measures 50 feet. Alana hit 4 home runs in a game. What was the total number of feet she ran in the game?

★12. Brian ran the perimeter of a field. He ran 150 feet to the east, turned 90° south, and ran 90 feet. He then turned west and ran 150 feet. He turned 90° north and ran 90 feet. How many feet did he run? What was the shape of the field?

8.5 Quadrilaterals 227

8.6 Problem-Solving Strategy: Make a Drawing

Objective: to make a drawing to solve a problem

On a sight-seeing trip to Washington, DC, Andy and Will left their hotel and walked 6 blocks west, then 4 blocks north, and 2 blocks east. Then they discovered they were lost. How can they get back to their hotel?

1. READ You need to find how to get back to the hotel. You know the directions and distances the boys have traveled.

2. PLAN You can make a drawing to find directions to the hotel.

3. SOLVE On grid paper draw the path the boys have taken from the hotel.

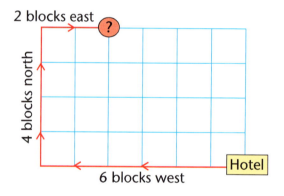

Since the boys are now 4 blocks west of the hotel, they should go 4 blocks east, and then 4 more blocks south to return to their hotel.

What other possible solutions are there to the problem?

4. CHECK Count the blocks in the drawing. If the boys go 4 blocks east and 4 blocks south they will return to the hotel.

TRY These

1. When Stephen delivers newspapers he leaves his house, travels 3 blocks east, 4 blocks south, and then 5 blocks west. How might he return home?

Solve

1. Tom's birthday is before Lee's. Judy's birthday is right after Tom's. In what order do they celebrate their birthdays?

2. If you plant a row of 12 beans by putting the seeds 5 inches apart, how far is the last bean from the first one?

3. The cafeteria serves a different kind of soup each day of the week. The vegetable soup is served earlier in the week than the mushroom, but later than the tomato. The mushroom soup is served the day before the broccoli, and the day after the chicken noodle. In what order is the soup served?

4. Mark packs cereal boxes into trays. If a box that contains one serving of cereal is 2 inches wide and 3 inches long, how many boxes can he fit into one layer of a tray that is 6 inches wide and 8 inches long? Use grid paper for your drawing.

5. If you borrow a book from the library on January 6, and it is due in 2 weeks, on what date is it due?

MID-CHAPTER Review

Draw an angle for each given measure.

1. 85°
2. 55°
3. 165°

Classify each triangle by its sides and angles.

4.
5.
6.

8.6 Problem-Solving Strategy: Make a Drawing

8.7 Circles

Objectives: to construct a circle and identify its parts; to find the diameter and radius of a circle

Engineers have designed pipelines to irrigate large land areas. One type works like a minute hand of a clock because it moves around a center point that is the source of the water. As the pipeline turns, the support wheel at the end draws a circle on the land.

A **circle** is a set of points that are the same distance from a given point. That point is the center of the circle. A circle has no beginning point and no ending point.

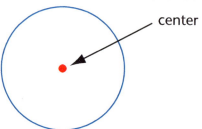
center

Parts of a Circle

A line segment that connects any two points on the circle is called a **chord**.

A chord that passes through the center of the circle is called a **diameter**.

A line segment that connects the center with a point on the circle is called a **radius**.

Exploration Exercise

A **compass** is a tool for constructing circles.

Directions:

1. Draw a point. Place the point of the compass over the point.
2. Set the compass to the length of the radius, 3 cm.
3. Hold the compass still with your finger at the center point, and move the compass to make the circle.
4. Draw and label a chord, a diameter, and a radius.
5. How are the chord and a diameter alike? How are they different?
6. What is the measurement of the radius and diameter? What relationship do you notice between a radius and a diameter of a circle?

The diameter is twice the length of the radius. The radius is one-half the length of the diameter.

Examples

A. Find the diameter.

Since the diameter is twice the length of the radius, you should multiply the radius by 2.

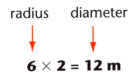

$6 \times 2 = 12$ m

The diameter is 12 m.

B. Find the radius.

Since the radius is one-half the length of the diameter, you should divide the diameter by 2.

diameter   radius

$15 \div 2 = 7\frac{1}{2}$ in.

The radius is $7\frac{1}{2}$ in.

TRY These

Use the diagram to the right to answer questions 1–4.

1. Name the center of the circle.
2. Name a diameter of the circle.
3. Name a chord in the circle.
4. Name a radius of the circle.

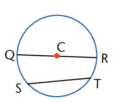

Exercises

1. Draw *three* circles of different sizes.
2. Measure each circle's radius and diameter.

8.7 Circles

Find the diameter of each circle below.

3.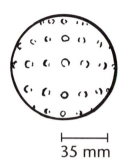
35 mm

4.
1.25 m

5.
$\frac{1}{8}$ yd

Find the radius of each circle below.

6.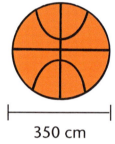
350 cm

7.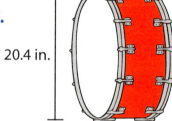
20.4 in.

8.
2.5 cm

PROBLEM Solving

9. A Ferris wheel has a diameter of 250 ft. Find the radius of the Ferris wheel.

10. If you know the length of a radius of a circle, how can you find the length of a diameter?

11. If you know the length of a diameter of a circle, how can you find the length of a radius?

12. Is every diameter a chord of a circle? Why or why not?

13. How many cuts that are diameters are needed to cut a pizza into eight pieces?

Constructed Response

14. Do two radii always form a diameter of a circle? Make a drawing to show your answer.

15. Is every chord a diameter of a circle? Why or why not?

Problem Solving

Pick a Pattern

A pentomino is a pattern made of five connected squares.
These patterns are pentominoes.

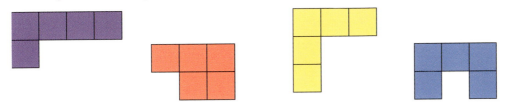

A flip or turn does not change the pentomino.
These are the same pentomino.

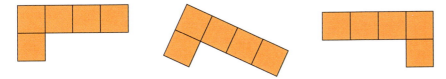

Draw *three* more pentominoes.
Can you find more?
How many different pentominoes are there?

Extension

Some of the pentominoes can be folded into open boxes. Guess which pentominoes will work. Then check by cutting and folding the pentominoes you have made.

8.8 Congruence

Objective: to identify and describe congruent shapes

Burt Blaney operates a press that forms the metal from which signs are made. All of the signs he makes have congruent shapes.

Congruent figures have the same size and shape. They will match or fit on each other exactly.

Examples

A. Line segments that have the same length are congruent to each other. If you trace $\overline{XY}$ and lay it on top of $\overline{RS}$, the two line segments match exactly.

B. Angles that have the same measure are congruent to each other. If you trace ∠B and lay it on top of ∠A, they match exactly.

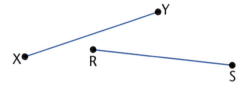

C. Triangle ABC will fit exactly on triangle DEF. The triangles are congruent. The parts that match are called **corresponding parts**. Name the sides that match. Name the angles that match.

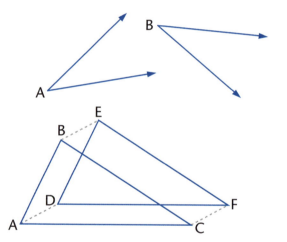

Name the letter of the figure that is congruent to the first figure.

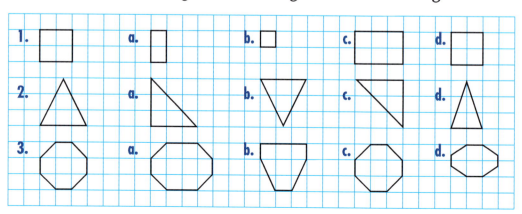

234   8.8 Congruence

Exercises

Use tracings to find the figure that is congruent to the first figure.

1.   a.   b.   c.

2. 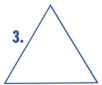   a.   b.   c.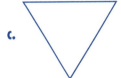

3.   a.   b.   c.

Problem Solving

4. Jack had an extra set of keys made for his car. Are the new keys congruent to each other? Why or why not?

5. The sides of figure A are congruent to the sides of figure B. Are figures A and B congruent? Why or why not?

6. Regular polygons have all sides and all angles congruent. Which of the polygons below are regular?

   a.   b.   c.

Constructed Response

7. One circle has a radius of 3 inches. Another circle has a diameter of 6 inches. Are the two circles congruent? Why or why not? Explain.

8.9 Symmetry

Objective: to identify line symmetry

Dave Dennison is a photographer. He has a collection of photographs that shows **symmetry** he observes in nature.

Certain figures and patterns have **lines of symmetry**. This means they can be folded along the line of symmetry and both sides match exactly.

Examples

Trace each figure. Then compare the figure across the line of symmetry. Name the polygon formed.

A.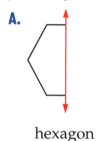
hexagon

B.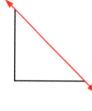
square

C.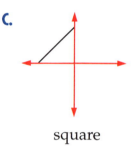
square

 Complete the figure across one line. Then complete the new figure across the second line.

More Examples

Use grid paper to copy each pattern. Then complete the pattern across the line of symmetry.

D.

E.

TRY These

Name the letter that appears when each letter card is unfolded.

1.
2.
3.
4.

Exercises

Trace each figure. Complete the figure across each line of symmetry.

1.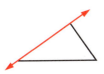
2.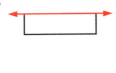
3.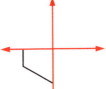
4.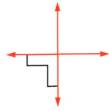

5.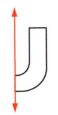
6.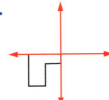
7.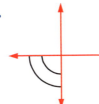
8.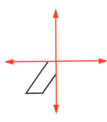

Use grid paper to copy each pattern. Then complete the pattern across the line of symmetry.

9.
10.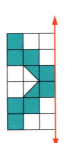
11.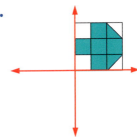

Tell how many lines of symmetry each figure has.

12.
13.
14.

Problem Solving

15. List the capital letters that have vertical line symmetry. Write words using the letters. Repeat using only letters with horizontal symmetry.

8.9 Symmetry 237

8.10 Transformations

Objective: to identify and describe the results of translations, reflections, and rotations of geometric figures

Sylvia Lee assembles circuit boards for use in computers. As she picks up a part to be installed on the board, she often has to turn it so it is in the right position for installation.

Figures can be moved in a plane in three ways. These movements are called **transformations.**

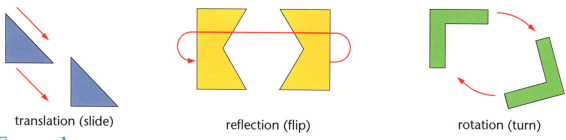

translation (slide)   reflection (flip)   rotation (turn)

Examples

Name the transformation needed to put the tangram piece in place.

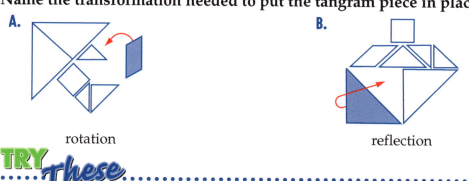

A. rotation   B. reflection

TRY These

State the transformation used to move each figure from position *a* to position *b*.

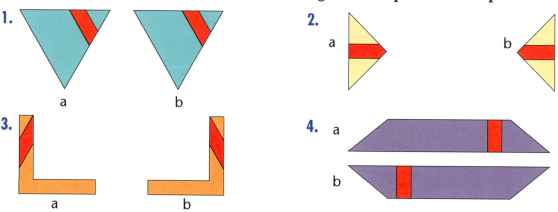

238   8.10 Transformations

Exercises

Name the transformation needed to move the piece to the dotted position.

1.

2.

3.

4.

Trace each figure. Then draw the figure in its new position after performing the given transformation.

5. reflection

6. translation

7. rotation

Problem Solving

8. What transformation is made when you turn the pages of a book?

9. What transformation is made when you move a dead-bolt lock?

10. Name some transformations in sports events that are examples of translations, reflections, and rotations.

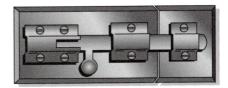

8.10 Transformations

8.11 Three-Dimensional Figures

Objective: to identify three-dimensional figures

Figures, such as buildings, that do not lie in a plane are called **three-dimensional figures**. Three-dimensional figures have special parts.

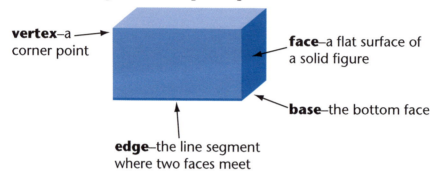

vertex—a corner point

face—a flat surface of a solid figure

base—the bottom face

edge—the line segment where two faces meet

There are many three-dimensional figures.

Rectangular prisms have two congruent rectangular bases and four rectangular faces.		**Triangular prisms** have two congruent triangular bases and three rectangular faces.	
Cubes have six congruent faces.		**Spheres** have one curved surface.	
Rectangular pyramids have one rectangular base and four triangular faces.		**Triangular pyramids** have one triangular base and three triangular faces.	
Cylinders have two circular bases. The bases are connected by a curved surface.		**Cones** have one circular base. A curved surface connects the base to the vertex.	

Examples

A. Name the figure.

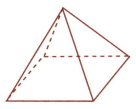

The base is a square, so it is a square pyramid.

B. Name the figure.

There are two circular bases connected by a curved surface, so it is a cylinder.

TRY These

Name each figure below.

1.
2.
3.

Exercises

Name each figure.

1.
2.
3.

4. Copy and complete the chart.

Three-Dimensional Figure	Number of Faces	Number of Edges	Number of Vertices
Rectangular prism			
Cube			
Square pyramid			
Sphere			

8.11 Three-Dimensional Figures

Sketch each three-dimensional figure.

5. triangular prism
6. cylinder
7. rectangular prism

8. Draw the figure that represents a textbook. Name the figure.
9. Draw the figure that represents a can of soup. Name the figure.

10. Which figure has exactly five faces and eight edges?

    a. square pyramid
    b. triangular prism
    c. rectangular prism
    d. triangular pyramid

11. Which figure has no flat surfaces?

    a. cone
    b. prism
    c. pyramid
    d. sphere

MIXED Review

Solve.

12. Mr. Blaney buys 60 octagonal tiles at $1.98 each and 72 square tiles at $1.69 each for a patio. What is the total cost of the patio tiles?

13. A new school desk costs $74.06. How much would 22 new desks cost?

14. The principal of Hughes Middle School ordered 625 boxes of chalk. If there are 25 boxes of chalk in each case, how many cases did she receive?

15. Mr. Allen drives to work, a distance of 17 miles. How far will he drive to and from work during a workweek of 5 days?

16. If Dennis multiplies his age by 6 and subtracts 15, he gets 153. How old is Dennis?

17. *USA Today* has a circulation of 1,591,629 newspapers. *The Baltimore Sun* has a circulation of 304,412. What is the difference in circulations?

Problem Solving

Puzzler

Skyman has been imprisoned by the Puzzler. To escape he must find the quickest way to move the tower of plutonium disks from one post to another so that the disks have the same arrangement as on the original post. He may move only one disk at a time. What is the minimum number of moves he must make in order to move the ten-disk tower and have it appear the same?

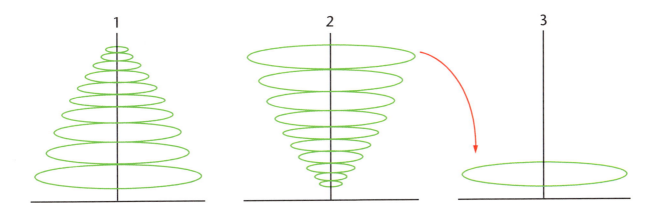

Chapter 8 Review

LANGUAGE and CONCEPTS

Choose the word in parentheses that makes each sentence true.

1. An endpoint and all the points that continue forever in a certain direction is called a (diameter, ray).
2. A polygon with five angles and five sides is called a (pentagon, hexagon).
3. A (ruler, protractor) can be used to draw or measure a 60° angle.
4. (Polygons, Circles) are named by the number of sides and angles.
5. One way figures can be moved in a plane is by a (vertex, translation).
6. A (trapezoid, square) has four right angles.
7. A (rectangle, rhombus) has four equal sides.

SKILLS and PROBLEM SOLVING

Tell whether each situation describes a *line*, a *line segment*, a *ray*, or an *angle*. (Section 8.1)

8. 50 yard line
9. corner of a picture frame
10. clock hands
11. pointer on a spinner

Name the kind of angle. Write *acute*, *right*, or *obtuse*. (Section 8.2)

12.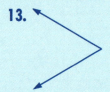
13.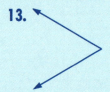
14.

Classify each triangle by its sides and angles. (Section 8.3)

15.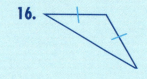
16.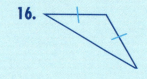
17.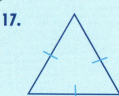

18. Two angles of a triangle have measures of 35° and 60°. What is the measure of the third angle?

Classify each polygon as a *triangle, quadrilateral, pentagon, hexagon,* **or** *octagon.* (Section 8.4)

19.   **20.** 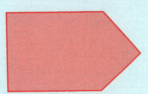 **21.**

Find the radius of each circle. (Section 8.7)

22. 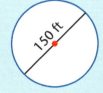 **23.**

Are the pairs of polygons congruent? Write *yes* **or** *no.* (Section 8.8)

24. 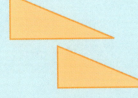 **25.**

Name the letter formed when the letter card is unfolded. (Section 8.9)

26.  **27.**  **28.**  **29.**

Solve. (Sections 8.6, 8.11)

30. Katie has math class before history. She has English right after math. In what order does she have her classes?

31. Draw a square pyramid. How many faces and edges are there?

Chapter 8 Review **245**

Chapter 8 Test

Use the figure at the right to answer questions 1–3.

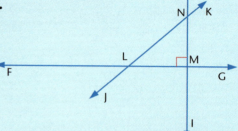

1. Name a set of perpendicular lines.
2. How many angles are there? Name them.
3. How many lines are there? Name them.

Draw angles for each given measure.

4. 45°
5. 140°
6. 30°
7. 105°

Classify each triangle in *two* ways.

8.
9.
10.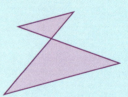

The measures of two angles of a triangle are given. Find the measure of the third angle.

11. 45°, 95°, _____
12. 30°, 60°, _____
13. 40°, 55°, _____

Tell whether each figure is a polygon. Write *yes* or *no*. If it is, name it.

14.
15.
16.

Write *yes* or *no* for questions 17–18.

17. A trapezoid has two pairs of parallel sides.

18. A rhombus has four congruent sides.

19. Trace the figure and complete it across each of the two given lines of symmetry.

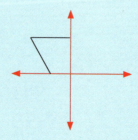

Change of Pace

Tessellations

When closed figures are arranged in a repeating pattern to cover a surface that has no gaps and no overlaps, it is called a **tessellation**. Tessellations can be found in designs used on walls, floors, and fabrics.

Look at the figure below. How do you know that the blue figure makes a tessellation?

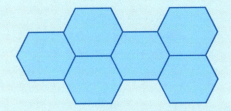

You can also make your own tessellation shapes by following the steps below.

Step 1	Step 2	Step 3	Step 4
Cut out a square.	Cut a notch out of the square.	Tape the notch to the opposite side.	Trace your design and tessellate.

Copy each figure. Write *yes* or *no* to tell whether each figure can be arranged to tessellate. If you can tessellate the figure, translate, reflect, or rotate it to make a design.

1.

2.

3.

Change of Pace 247

Cumulative Test

1. If 7 fair tickets cost $17.50, how much does 1 ticket cost?
   a. $2.50   b. $5.50
   c. $5.00   d. $7.50

2. 12,327 × 5 = _____
   a. 60,535
   b. 60,605
   c. 61,635
   d. none of the above

3. 1,460 ÷ 52 = _____
   a. 18 R44
   b. 26 R4
   c. 28 R4
   d. 38 R4

4. Bobbie purchased 6 dozen cookies for her party. She received $7.52 change from $20.00. How much did each dozen cost?
   a. $1.25
   b. $2.08
   c. $3.33
   d. $12.48

5. During 1 hour, 64 children and 38 adults visited the zoo. How many people visited the zoo that hour?
   a. 26 people
   b. 36 people
   c. 92 people
   d. 102 people

6. Find the best estimate for the measure of the angle.

   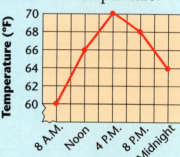

   a. 90°
   b. 180°
   c. 30°
   d. 120°

7. Which shows congruent figures?

8. Which is the difference between the 8:00 A.M. and 8:00 P.M. temperature?
   a. 4°F
   b. 8°F
   c. 60°F
   d. 68°F

9. Jack bought 5 bananas and 6 oranges. How much change did he receive from $5.00? What information is needed to solve the problem?
   a. how much money Jack took to the store
   b. the cost of 1 banana and 1 orange
   c. the difference in cost between bananas and oranges
   d. none of the above

10. Which of the following shows a ray?

    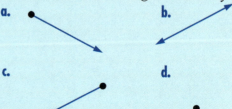

CHAPTER 9

Perimeter, Area, and Volume

Megan DeBoard
Honolulu, HI

9.1 Perimeter of Polygons

Objective: to find the perimeter of polygons

What is the length of a bicycle track that borders the edge of the golf course shown to the right?

The distance around a polygon is called the **perimeter**. You can add the measures of the sides of a polygon to find the perimeter.

side + side + side + side = perimeter
0.6 + 0.5 + 0.9 + 0.4 = 2.4

The golf course is shaped like a polygon.

The length of the bicycle track is 2.4 miles.

More Examples

You can find the perimeter of some polygons in different ways.

A. If all sides of this pentagon have the same length, you can multiply instead of adding.

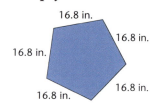

5 × 16.8 = 84
The perimeter is 84 in.

B. Some polygons have pairs of sides that have equal lengths.

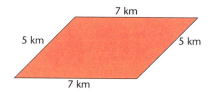

(2 × 7) + (2 × 5)
   14   +   10   = 24
The perimeter is 24 km.

TRY These

Find the perimeter of each polygon.

1. Rectangle with sides 3 m, 7 m, 3 m, 7 m

2. Square with sides 25 in.

3. Triangle with sides 16 cm, 22 cm, 20 cm

Exercises

Find the perimeter of each polygon.

1.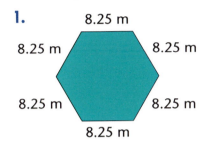

2.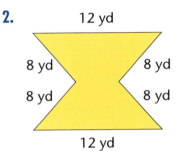

3.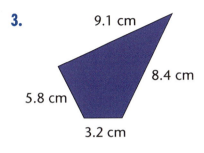

Find the perimeter of a polygon with sides of the given lengths.

4. $\frac{1}{3}$ mi, $\frac{1}{4}$ mi, $\frac{1}{6}$ mi

5. 2.6 m, 5 m, 20.3 m, 12.5 m

Problem Solving

6. A rectangle is 6 inches wide and 11 inches long. Find the perimeter.

7. Draw a square with a perimeter of 20 centimeters. Label the length of one side.

★8. How many different rectangles having a perimeter of 24 feet can be drawn if the side measures are whole numbers? Draw the rectangles on grid paper.

Mind Builder

Perimeters

Trace and cut out the figures below.

   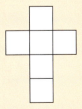

Arrange the figures to form one polygon. The perimeter of the polygon should be 26 units less than the sum of the perimeters of the single figures.

9.2 Circumference

Objective: to use pi to find the circumference of circles

Mr. Stoddard wants to put a metal strip around his archery target. The diameter of the target is 48 inches. How many inches of metal strip does Mr. Stoddard need?

You need to find the circumference of the target, but before you can do that, you need to know about a special number called **pi**. The boundary or **circumference** of any circle is the distance across it times pi.

Pi is roughly 3.14. That means that for every circle the circumference is about 3.14 times the diameter. The Greek letter pi (π) represents pi.

> **circumference ÷ diameter = π or $C \div d = \pi$**

> Remember the Commutative Property.
> If $6 \div 3 = 2$,
> then $2 \times 3 = 6$,
> or $6 = 2 \times 3$.

The related multiplication equation can help you find the circumference.

> **circumference = $\pi \times$ diameter or $C = \pi \times d$**

$C = 3.14 \times 48$ or 150.72 inches.

Round to the nearest whole number that makes sense. Mr. Stoddard needs to buy at least 151 inches of metal strip.

Another Example

Find the circumference of a circle if the radius is 25 yards.

$C = \pi \times d$
$C = 3.14 \times 50$ yd

The radius is half the diameter, so the diameter is 2×25 or 50 yd.

The circumference of the circle is 157 yards.

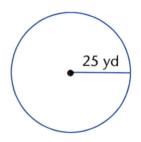

TRY These

Name the radius and diameter for each circle.

1. radius
   ■ ft
   diameter
   ■ ft

2. radius
   ■ m
   diameter
   ■ m

3. radius
   ■ in.
   diameter
   ■ in.

Name the diameter for each radius given. Then estimate the circumference. Use 3 as an estimate for pi.

1. 8 miles
2. 11 meters
3. 91 centimeters
4. 6.2 feet

Find the circumference of each circle. Use 3.14 for π.

5.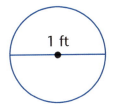
6.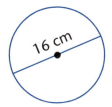
7.
8.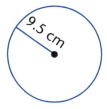

9. The circumference of a circle is 28.26 feet. Using 3 for pi, find the approximate diameter.

10. The circumference of a circle is 18.84 meters. Using 3 for pi, find the approximate radius.

11. You measure the outside of a circular pool and find the circumference to be 47 feet. About how long is one lap across the center of the pool?

★12. How much longer is the perimeter of the square than the circumference of the circle?

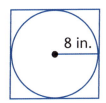

Constructed Response

13. Mrs. Stoddard wants to have a round tabletop made from an oak tree in her yard. If the tree is 10 feet around, about how long could be the diameter of her table? Draw a diagram and explain.

Estimate.

14. 3.7 × 5.2
15. 9.1 × 2.2
16. 4.6 × 8.3
17. 1.8 × 6.5
18. 9.6 × 1.1
19. 10.3 × 9.1
20. 11.4 × 4.1
21. 12.6 × 1.1

9.2 Circumference 253

9.3 Area of Rectangles

Objective: to find the area of rectangles

The amount of surface an object covers is called its **area**. The area of a shape is the number of units needed to cover the shape. It might be the amount of carpet needed to cover a floor, or the amount of sod to cover a football field.

You can count to find the area of this rectangle. Each square is 1 centimeter long and 1 centimeter wide or 1 square centimeter (cm^2).

The area of the rectangle is 12 cm^2. It is important to give area answers in square units.

Some common units of area are listed below. Which unit of area do you think would be used to measure a football field?

square millimeter (mm^2)   **square meter (m^2)**   **square kilometer (km^2)**
square inch ($in.^2$)   **square foot (ft^2)**   **square yard (yd^2)**

Another Example

To find the area (A) of a rectangle, multiply the length (ℓ) by the width (w).

Area is equal to length times width.

$A = \ell \times w$
$A = 3.4 \times 1.2$
$A = 4.08$

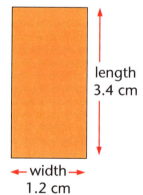

length 3.4 cm

width 1.2 cm

The area is 4.08 square centimeters (4.08 cm^2).

TRY These

Multiply to find the area of each rectangle.

1.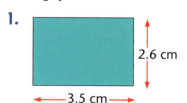

2.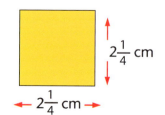

Exercises

Find the area of each rectangle.

1.
2.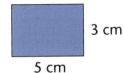
3.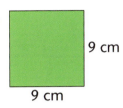

4.
5.
6.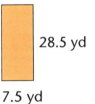

Find the area of each rectangle described below.

7. width 6 m; length 7 m
8. width 5 cm; length 8 cm
9. width 13 ft; length 15 ft
★ 10. width 6 ft; length 3 yd

Problem Solving

11. A regulation tennis court is 24 meters long and 11 meters wide. How many square meters is it?

12. A rectangle is 4 meters wide. Its area is 32 m². What is the length of this rectangle?

13. The library needs new carpet that is usually sold by the square yard. If the library is 12 yards by 15 yards, how much carpet should be ordered?

★ 14. Sod is sold by the square foot. How much sod is needed for a football field that is 120 yards long and 160 feet wide?

9.4 Perimeter and Area of Complex Figures

Objective: to find the perimeter and area of complex figures

Martha's vegetable garden is pictured to the right. Martha wants to find the perimeter and area of the garden.

Finding the Perimeter of a Complex Figure

Step 1	Step 2
Find any missing lengths. You know the opposite sides of a rectangle are equal. To find the missing side subtract. 6 ft – 2 ft = 4 ft The missing side is 4 ft.	Add the lengths of all the sides. 2 ft + 6 ft + 5 ft + 2 ft + 3 ft + 4 ft = 22 ft The perimeter is 22 ft.

Finding the Area of a Complex Figure

Step 1	Step 2	Step 3
Divide the complex figure into simple figures.	Use the formula to find the area of each simple figure. $Area_1 = 2 \text{ ft} \times 6 \text{ ft} = 12 \text{ ft}^2$ $Area_2 = 2 \text{ ft} \times 3 \text{ ft} = 6 \text{ ft}^2$	Add the areas. Area of $rectangle_1 = 12 \text{ ft}^2$ Area of $rectangle_2 = 6 \text{ ft}^2$ Area of complex figure $= 12 \text{ ft}^2 + 6 \text{ ft}^2 = 18 \text{ ft}^2$

TRY These

Find the lengths of the missing sides in each complex figure.

1.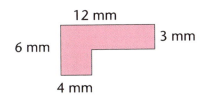

2.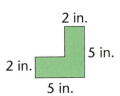

Exercises

Find the perimeter and area of each complex figure.

1.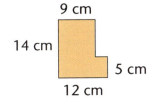

2.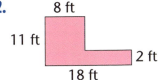

3.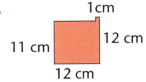

4.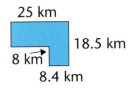

5.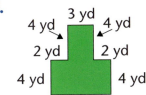

6.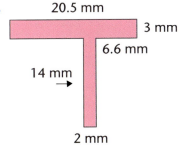

PROBLEM Solving

7. Draw *two* complex figures each with a perimeter of 25 centimeters.

8. Draw a complex figure on grid paper. Then find the perimeter and area of your figure.

TEST Prep

9. Find the area of this figure.

   a. 28.1 in.²
   b. 41.83 in.²
   c. 42.3 in.²
   d. 47.91 in.²

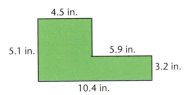

9.4 Perimeter and Area of Complex Figures   **257**

9.5 Problem-Solving Strategy: Draw it

Objective: to solve problems by making a drawing

Each of the nine members on Amanda's softball team likes to pitch. They decide to play "One Potato, Two Potato" to choose the pitcher. In which place in line should Amanda stand so she will be chosen?

1. READ You need to find out in which place in line Amanda should stand so that she will be the pitcher. You know there are 9 players, and that they will play "One Potato, Two Potato" to decide.

2. PLAN Draw and label pictures of the nine players.

3. SOLVE Say *one potato* at the first picture. At the next picture say *two potato* and so on. The picture who has more is out. Continue counting until only one picture is left. In which place in line should Amanda stand so that she can pitch today?

4. CHECK Move Amanda's picture to the place in line where she would be chosen. Repeat "One Potato, Two Potato."

TRY These

1. What is the least number of links you need to cut and glue back together to make one 15-link chain?

2. Trace pieces **A**, **B**, **C**, and **D**. Cut out the pieces and use them to form a square.

   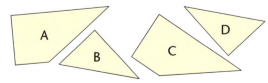

1. Make the letters form the words *TOP STAR*. Move only one letter at a time to the open space. The moves may be horizontal or vertical.

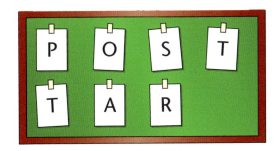

Solve.

2. Kirsten swam for $\frac{2}{3}$ hour in the morning and $\frac{1}{2}$ hour in the afternoon. How much longer did she swim in the morning? How many minutes is that?

3. Mr. Patton made a sandwich that weighed 3 pounds and was 1 yard long. How many ounces were left if the people at the party ate 2 pounds 4 ounces of it?

4. How far would you walk if you walked around the edge of the pool?

5. The bottom of the pool is green and the sides are white. Find the area of the pool that is green.

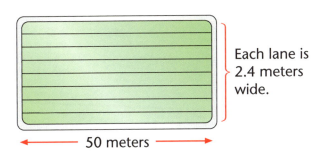

Each lane is 2.4 meters wide.

50 meters

9.6 Area of Parallelograms

Objective: to find and use the formula for the area of a parallelogram

After school, Paul and his friends play soccer on a vacant lot that is shaped like a parallelogram. How can you find the area of a parallelogram?

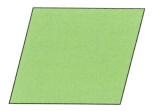

A **parallelogram** has two pairs of parallel sides.

Exploration Exercise

1. On grid paper, draw the parallelogram at the right. Then cut it out.

   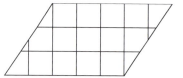

2. Estimate its area in square units.

3. The bottom edge of the parallelogram is called the **base**. How many units long is the base?

   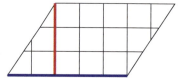

4. Draw a segment from the top edge to the bottom edge along a grid line. The length of this segment is the **height** of the parallelogram. What is the height of this parallelogram?

5. Cut along the segment. Move the triangle that is formed to the other end of the parallelogram.

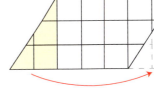

6. What kind of figure have you formed?
   What is the area of this figure?
   Is the area of this figure the same as the area of the parallelogram?

7. What is the area of the parallelogram?
   How does your answer compare to your estimate?
   Compare the area to the base and height. What do you notice?

To find the area (A) of a parallelogram, multiply the base (b) by the height (h).

Example

Find the area of the parallelogram.

$A = b \times h$
$A = 5 \times 2$
$A = 10$

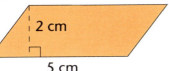

A right angle is formed by the base and the height.

The area of the parallelogram is 10 cm².

TRY These

Find the area of each parallelogram.

1.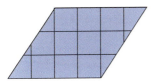

2.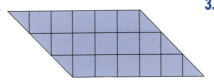

3.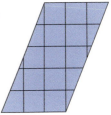

Identify the base and height of each parallelogram. Then find the area of each.

4.

5.

6.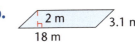

9.6 Area of Parallelograms 261

Exercises

Find the area of each figure.

1.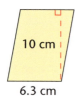

2.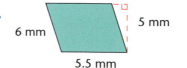

3.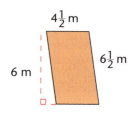

4. Draw and label a model of a parallelogram with a base of 9 feet and an area of 36 square feet.

PROBLEM Solving

5. A piece of property shaped like a parallelogram has a base of 62 feet and a height of 27 feet. How many square feet is it?

★ 6. Look at the rectangle and parallelogram below. Explain how you can show a classmate that they have the same area.

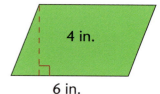

Constructed Response

7. Mr. Kwan wants to paint a 4 ft by 8 ft sheet of plywood on one side. Does he have enough paint? Explain.

MIXED Review

Compare using <, >, or =.

8. 0.6 ● 0.9

9. 0.35 ● 0.350

10. 3.21 ● 3.12

Order from least to greatest.

11. 0.9   0.8   1.1

12. 1.46   1.64   1.06

Problem Solving

The Great Divide

Six students share a dormitory suite. Read their requests below. Then match the students with their dorm rooms.

1. Connie, Peg, and Rosa want their desks in the study area.
2. Peg wants to be next to Yvonne and Justine.
3. Margaret wants to be near the hallway.
4. Yvonne and Connie want corner rooms.

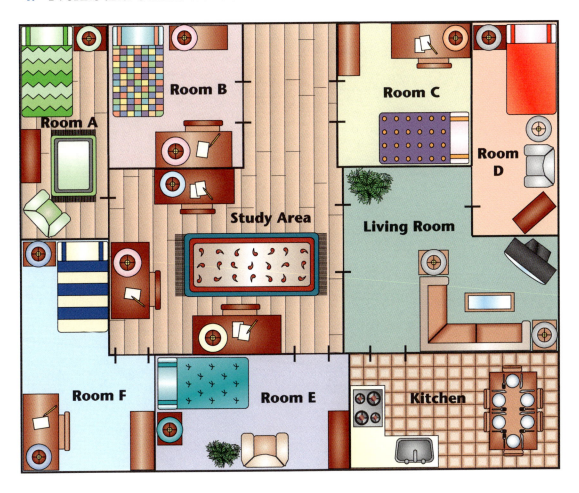

9.7 Area of Triangles

Objective: to find and use the formula for the area of a triangle

Exploration Exercise

1. On grid paper, draw the parallelogram at the right. Then cut it out.

2. Find the base, height, and area.

3. Draw a segment from one corner to the opposite corner. Cut along the segment.

4. What kind of figures do you have now?

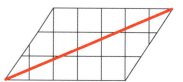

5. What is the area of each figure? How do you know?

To find the area (A) of a triangle, find $\frac{1}{2}$ of the base (b) times the height (h).

$A = \frac{1}{2} \times b \times h$

$A = \frac{1}{2} \times 6 \times 2$ Replace *b* with 6 and *h* with 2.

$A = 3 \times 2$

$A = 6$

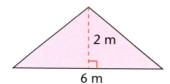

A right angle is formed by the height and base.

The area of the triangle is 6 m².

Another Example

Find the area of the triangle.

$A = \frac{1}{2} \times b \times h$

$A = \frac{1}{2} \times 6 \times 12$

$A = 3 \times 12$

$A = 36$

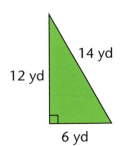

In a right triangle, the height is one of the sides that make the right angle.

The area is 36 yd².

TRY These

Identify the base and height of each triangle. Then find the area of each.

1.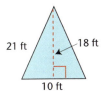
2.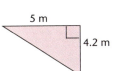
3.

Exercises

Find the area of each.

1.
2.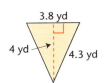
3.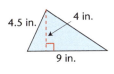

4. Draw and label a model of a triangle with a base of 3 meters and a height of 10 meters. Find the area.

9.7 Area of Triangles

PROBLEM Solving

5. A triangular park has a base of 180 feet and a height of 86 feet. What is its area?

6. What is the area of the unshaded part of the square? Describe how you found the area.

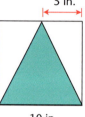

★7. A right triangle has an area of 20 cm². If the base measures 10 cm, what is the height?

Constructed Response

8. How can knowing the area of a parallelogram help you find the area of a triangle? Give an example.

MID-CHAPTER Review

Find the perimeter and area of each figure.

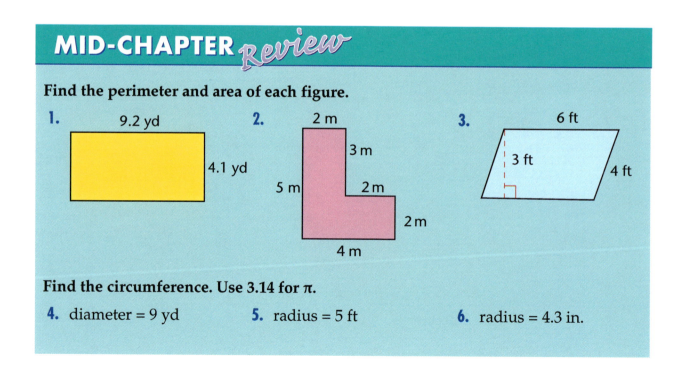

Find the circumference. Use 3.14 for π.

4. diameter = 9 yd
5. radius = 5 ft
6. radius = 4.3 in.

9.7 Area of Triangles

Cumulative Review

Compute.

1. 12,836 + 42,963
2. 43.7 + 17.4
3. 7.306 + 9.470
4. $33.52 − 6.89
5. 18,472 − 4,003
6. 96,000 − 47,372
7. 2,370 × 201
8. 34.3 × 4.2
9. 8)2,344
10. 4)8.016
11. 5)15.6
12. 9)210.87

Write each improper fraction as a whole number or mixed number in simplest form.

13. $\frac{9}{5}$
14. $\frac{12}{8}$
15. $\frac{16}{4}$
16. $\frac{21}{12}$
17. $\frac{9}{2}$
18. $\frac{18}{4}$

Compute. Write each answer in simplest form.

19. $\frac{5}{8} + \frac{1}{8}$
20. $\frac{3}{5} + \frac{1}{10}$
21. $\frac{5}{6} - \frac{1}{6}$
22. $\frac{9}{7} - \frac{2}{7}$
23. $\frac{2}{3} + \frac{1}{6}$
24. $\frac{3}{8} + \frac{1}{2}$
25. $\frac{7}{8} - \frac{1}{4}$
26. $\frac{2}{3} - \frac{5}{9}$

Compare using <, >, or =.

27. 2.8 ● 2.80
28. 32.6 ● 3.28
29. 0.6 ● 0.52
30. $\frac{2}{3}$ ● $\frac{5}{6}$
31. $\frac{1}{2}$ ● $\frac{3}{4}$
32. $\frac{1}{4}$ ● $\frac{1}{6}$

Solve.

33. A rectangle is 12.8 inches long and 7.4 inches wide. What is the perimeter?

34. A square has an area of 64 m². How many centimeters long is each side of the square?

35. A square field is 50 meters on a side. How much fencing is needed to enclose the field?

Cumulative Review

9.8 Problem-Solving Strategy: Using Venn Diagrams

Objective: to solve word problems using Venn diagrams

For her health project, Karen surveyed the students in her class to find out if they like green beans, peas, or both. She found out that 16 students like green beans, 13 like peas, and 6 like both vegetables. All of the students, including Karen, like at least one vegetable. How many students are in Karen's class?

Names	Green Beans	Peas
Mark	X	
Karen	X	X
Sheila		X
Dave	X	
Tai	X	X

1. READ — You know the number of students who like each kind of vegetable, and the number of students who like both vegetables.

You need to find how many students are in Karen's class.

2. PLAN — Use a **Venn diagram** to show what you know. A rectangle represents all the students. A circle represents students who like one kind of vegetable. Where the circles overlap shows the students who like both green beans and peas.

3. SOLVE

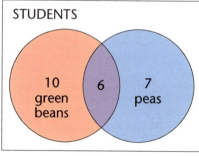

Start with the students who like both vegetables. Write a 6 in the area where the circles overlap.

Subtract 6 from 16 to find out how many like green beans but do not like peas. Likewise, subtract 6 from 13 to find how many like peas and not green beans.

Add the numbers in each part of the circles to find the total number of students.

There are 23 students in Karen's class. $10 + 6 + 7 = 23$

4. CHECK

Students who like green beans		Students who like peas		Students who like both		Total number of students
16	+	13	−	6	=	23

There must be less than 29 students since 6 of the students like both vegetables.   **23 < 29**   The answer makes sense.

Use the Venn diagram of Karen's class to solve.

1. How many students like to ride bicycles?
2. How many students like only swimming?
3. How many students like to ride bicycles and swim?
4. If there are 23 students in Karen's class, how many students like neither swimming nor riding bicycles?

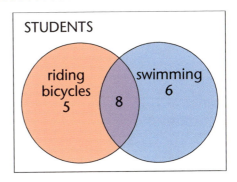

Draw a Venn diagram.

1. For lunch, 12 of Karen's classmates ate cookies and 16 ate melon. How many of the 23 students ate both cookies and melon?

2. Of the 15 sandwiches made for a picnic, 10 had ham and 3 of those had both ham and cheese. How many sandwiches had only cheese?

Use any strategy.

3. What is the price of each orange? 18 oranges?

5. Each library shelf holds 25 books. How many shelves are needed to hold 750 books?

7. Kate's recipe for vegetable soup calls for 5 cups of water for every 2 cups of vegetables. To make soup for 6 cups of vegetables, how much water should she use?

Use the data from the previous page.

4. What part of Karen's class likes both green beans and peas? Write your answer as a fraction.

6. If Mike uses 2 eggs to make 6 waffles, how many eggs does he need to make 24 waffles?

★8. Patti can make 3 belts in 45 minutes. How many minutes will it take her to make 5 belts?

9.8 Problem-Solving Strategy: Using Venn Diagrams   269

9.9 Three-Dimensional Figures and Nets

Objective: to recognize a two-dimensional pattern for a three-dimensional solid

A **net** is a two-dimensional pattern for a three-dimensional solid. If you cut a prism along some of its edges and lay it out flat, you have a net of the prism. Sometimes you may find that there is more than one possible net for a figure.

Look at the net to the right. Think of different ways that you can fold it to make a solid figure.

How many faces do you see?

What is the shape of the base? What are the shapes of the other faces?

What three-dimensional figure can you make by folding the net?

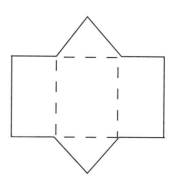

TRY These

Match each three-dimensional figure with its net. Write *a*, *b*, *c*, or *d*.

1.

2.

3.

4.

a.

b.

c.

d.

Exercises

Match each three-dimensional figure with its net. Write *a*, *b*, *c*, or *d*.

1.   a.

2.   b.

3.   c.

4.   d.

PROBLEM Solving

5. Cristina is making nets of 5 cubes for a project. How many faces will she draw?

6. Draw a net for a cone.

7. Draw a net for a rectangular pyramid.

8. Draw a net for a cylinder.

TEST Prep

9. Kate wants to make a triangular prism using a set of plane figures. Which set can she use?

   a. 2 triangles and 3 rectangles
   b. 1 triangle and 4 rectangles
   c. 4 triangles and 1 rectangle
   d. 3 triangles and 1 rectangle

9.9 Three-Dimensional Figures and Nets

9.10 Surface Area

Objective: to find the surface area of rectangular prisms and cubes

Surface area is the total area of the faces of a solid. The rectangular prism shown has six rectangular faces.

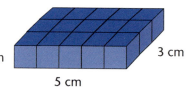

If you cut the rectangular prism along its edges and unfolded it, the net would look like this.

You can count all of the square units to find the surface area. You can also find the area of each face and add the areas.

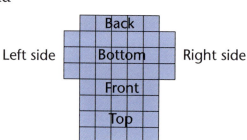

Step 1	Step 2	Step 3
Identify the faces.	Find the area of each face.	Add the areas of the faces to find the surface area.
The rectangular prism has six faces: top, bottom, left side, right side, front, and back.	Top = 3 × 5 = 15 cm² Bottom = 3 × 5 = 15 cm² Left side = 3 × 1 = 3 cm² Right side = 3 × 1 = 3 cm² Front = 5 × 1 = 5 cm² Back = 5 × 1 = 5 cm²	15 + 15 + 3 + 3 + 5 + 5 = 46 cm²

The surface area of the rectangular prism is 46 cm².

Another Example

Find the surface area of the cube.

272   9.10 Surface Area

Each face of a cube has the same area.

Area of a face = 2.5 in. × 2.5 in. = 6.25 in.^{2} Find the area of one face.
Surface area of the cube = 6.25 in.^{2} × 6 = 37.5 in.^{2} Multiply by 6.
The surface area of the cube is 37.5 in.^{2}

TRY These

On grid paper, draw a net for each prism. Then find the surface area.

1. 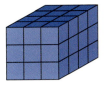   2.    3.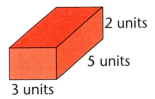
2 units
5 units
3 units

Exercises

Find the surface area of each prism.

1.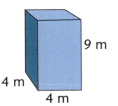
9 m
4 m
4 m

2.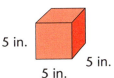
5 in.
5 in.
5 in.

3.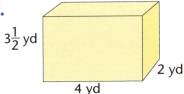
$3\frac{1}{2}$ yd
2 yd
4 yd

4.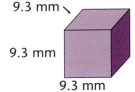
9.3 mm
9.3 mm
9.3 mm

5.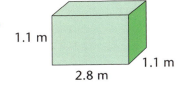
1.1 m
1.1 m
2.8 m

PROBLEM Solving

6. What is the total surface area of 5 cubes if each cube has 1-cm sides?

★7. If 27 1-cm cubes are put together to make a large cube, what is the surface area of the large cube?

9.10 Surface Area **273**

9.11 Volume

Objective: to find the volume of rectangular prisms and cubes

The box is shaped like a rectangular prism.

To find the volume (V) of any rectangular prism, multiply the length (ℓ) by the width (w) by the height (h).

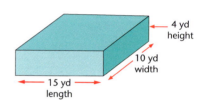

$V = \ell \times w \times h$ The length times the width equals the area of the base.
$V = 15 \times 10 \times 4$
$V = 150 \times 4$ The area of the base times the height equals the volume of the rectangular prism.
$V = 600$

The volume of the box is 600 cubic yards (yd^3).

For volume you use cubic units of measurement, such as 1 cubic meter, 1 cubic foot, or 1 cubic yard. It can be written as cu yd or yd^3.

The chart lists several common units of volume.

How can you remember that a cubic meter is written as m^3 instead of m^2?

Unit of Length	Unit of Volume
millimeter	cubic millimeter (mm^3)
centimeter	cubic centimeter (cm^3)
meter	cubic meter (m^3)
inch	cubic inch ($in.^3$)
foot	cubic foot (ft^3)
yard	cubic yard (yd^3)

TRY These

Find the volume of each rectangular prism.

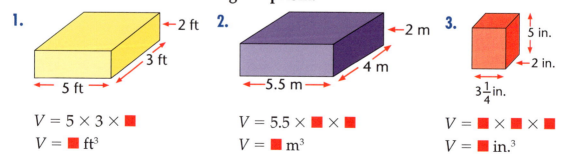

1. $V = 5 \times 3 \times \blacksquare$
    $V = \blacksquare \ ft^3$

2. $V = 5.5 \times \blacksquare \times \blacksquare$
    $V = \blacksquare \ m^3$

3. $V = \blacksquare \times \blacksquare \times \blacksquare$
    $V = \blacksquare \ in.^3$

Exercises

Find the volume of each rectangular prism.

1.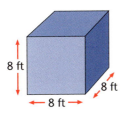

2.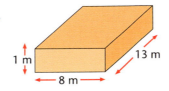

3.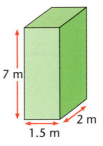

4.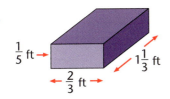

5.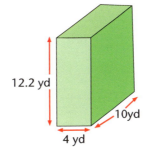

6.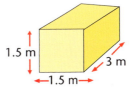

PROBLEM Solving

7. A refrigerator's dimensions are 6 feet by 4 feet by 4 feet. How many cubic feet of space does it occupy?

8. If a kitchen cabinet is 3 feet high, 2 feet deep, and 2 feet wide, how many cubic feet is it?

★ 9. Each edge of the cube measures 20 mm. Into how many 10-mm cubes can it be cut?

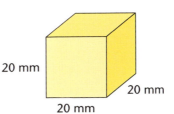

MIXED Review

Multiply or divide.

10. 4.36 × 5

11. 174 × 3.2

12. 9.16 × 5.2

13. 0.5 × 0.8

14. 0.06 × 0.9

15. 6)7.8

16. 9)75.6

17. 4)7.632

18. 5)340.5

19. 7)32.34

9.11 Volume 275

9.12 Problem-Solving Application: Using Perimeter, Area, or Volume

Objective: to use perimeter, area, or volume to solve problems

Louise and Pete are playing tennis with Russ and Kate. Since they are playing doubles they use the whole court. How much playing space do they have?

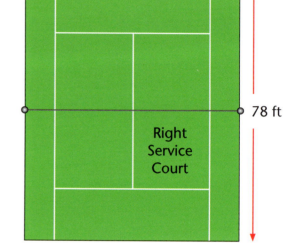

1. READ — You need to find how much playing space the four players have. You know they use the whole court. You know from the drawing that the length of the court is 78 feet, and the width of the court is 36 feet.

2. PLAN — To find the playing space or area, multiply the length by the width.

3. SOLVE — Multiply 78 by 36.   Estimate. 80 × 40 = 3,200

78 × 36 = 2,808

The playing area of the court is 2,808 square feet.

4. CHECK — Compared to the estimate, the answer is reasonable.

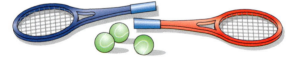

Which measurement do you need to find? Write *perimeter, area,* or *volume*.

1. You want to find the number of cubic feet of concrete used to make the tennis court.

2. You want to find the number of feet of fencing needed to enclose the tennis court.

3. Wilda painted the outer boundary lines of the court. What is the total length of the lines she painted?

4. When it is Gil's turn to serve, he must hit the ball into the right service court. How much space does he have onto which to hit the ball?

1. **Use the data from the previous page.** Pete and Russ decide to play a game of tennis. The width of the court for two players is 9 feet less than for doubles, but the length is still 78 feet. How much playing space do they have?

2. If a locker door measures 9 inches wide by 60 inches high, and the locker is 16 inches deep, what is the locker's capacity?

3. Virginia wants her grandmother to put lace around a napkin. If each side of the napkin is 7 inches, how much lace will her grandmother use?

4. Which storage compartment below has more storage space? how much more?

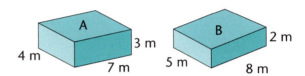

Solve.

5. Russ's tennis racket weighs 14 ounces. Kate's racket weighs one ounce less. What is the total weight of the rackets in pounds?

6. If 3 tennis balls cost $2.50, how much does a dozen balls cost?

7. Russ and Pete started playing tennis at 9:15 A.M. It took them 1 hour and 50 minutes to finish a match. At what time did they finish?

8. Kate put a round tablecloth on a picnic table near the tennis court. The tablecloth used has a diameter of 42 inches. What is its circumference?

Divide.

9. $2\overline{)82}$

10. $3\overline{)63}$

11. $4\overline{)56}$

12. $4\overline{)920}$

13. $2\overline{)127}$

14. $5\overline{)357}$

15. $8\overline{)4,046}$

16. $7\overline{)73,484}$

9.12 Problem-Solving Application: Using Perimeter, Area, or Volume

Chapter 9 Review

LANGUAGE and CONCEPTS

Write the word that best completes the sentence.

1. The distance around a square is the perimeter and the distance around a circle is the _____.

2. Area is measured in square feet, but _____ is measured in cubic feet.

3. A centimeter cubed equals 1 cm³, and a centimeter _____ is measured as 1 cm².

4. The area of a _____ is half the area of a parallelogram with the same base and height.

5. Circumference divided by diameter equals _____.

6. $A = b \times h$ is the formula for the area of a _____.

parallelogram
squared
pi
volume
triangle
circumference

SKILLS and PROBLEM SOLVING

Find the perimeter of each polygon. (Sections 9.1, 9.4)

7.

8.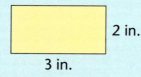

9.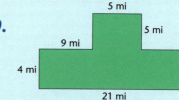

Find the circumference of each circle. (Use 3.14 for π. Section 9.2)

10.

11.

12.

Find the area of each figure. (Sections 9.3–9.4, 9.6–9.7)

13.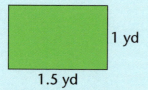

14.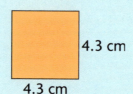

15.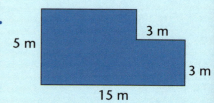

278 Chapter 9 Review

16. 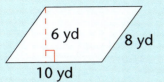   17. 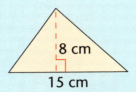   18.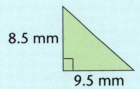

Predict what shape each net will make. (Section 9.9)

19. 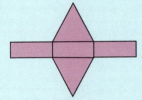   20. 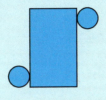   21.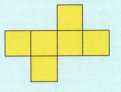

Determine the surface area and volume of each figure. (Sections 9.10–9.11)

22. 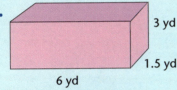   23.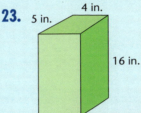

24. 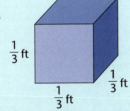   25.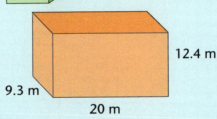

Solve. (Sections 9.5, 9.8, 9.12)

26. The perimeter of a square is 60 cm. What is the length of each side? What is the area of the square?

27. A large fish tank is 1 m by 0.4 m by 0.5 m. What is the volume of the tank?

28. Marc was the fifth person in the lunch line. Tom was two places in front of him and Rosemary was twelve places behind Tom. What place in line does Rosemary have?

29. In Jonathon's class, there are 25 students. Of the students 20 like the Ravens and 22 like the Orioles. If every students likes at least one of the teams, how many like both teams?

Chapter 9 Review 279

Chapter 9 Test

Find the perimeter of each figure.

1.
2.
3.

Find the circumference of each circle. Use 3.14 for π.

4.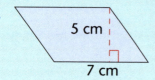
5.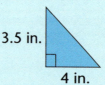
6.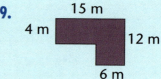

Find the area of each figure.

7.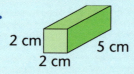
8.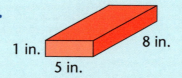
9.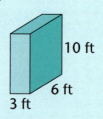

Find the surface area and volume of each rectangular prism.

10.
11.
12.

Solve.

13. The volume of the box below is 32 cubic feet. What is the height of the box?

14. Susie did a survey to find out if more students liked vanilla or chocolate ice cream. She found that 30 students liked vanilla, 20 students liked chocolate, and 12 liked both. If every student liked at least one kind, how many students were asked?

Change of Pace

Drawing Figures and Perimeter

Figures A and B have the same area (20 square units). The perimeter of each figure is different.

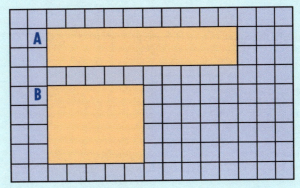

Figure A: perimeter = 24 units

Figure B: perimeter = 18 units

Copy and complete the chart. Draw each rectangle on grid paper.

	Length (units)	Width (units)	Perimeter (units)	Area (square units)
1.	24	1		
2.	12	2		
3.	8	3		
4.	6	4		

5. Describe the figure with the least perimeter.

6. Describe the figure with the greatest perimeter.

Solve.

7. Rosie wants to build a dog pen with an area of 16 square meters. What is the least amount of fencing she will need to enclose the pen?

8. Ben wants to plant a border of flowers around his vegetable garden. What shape will need the least amount of flowers, a 5 × 7 rectangle or a 5 × 5 square?

Cumulative Test

1. 30 ÷ 40 = _____
   a. 0.75
   b. 1.1
   c. 1.33
   d. none of the above

2. How many grams are there in 0.2 kilograms?
   a. 20 g
   b. 200 g
   c. 2,000 g
   d. 20,000 g

3. Ms. Houser's boat weighs 2 tons. How many pounds does the boat weigh?
   a. 1,500 lb
   b. 3,000 lb
   c. 4,000 lb
   d. 6,000 lb

4. Three yards is the same as how many inches?
   a. 36 in.
   b. 108 in.
   c. 180 in.
   d. 432 in.

5. Find the area of the right triangle.
   a. 2,048 cm^2
   b. 204.8 cm^2
   c. 20.48 cm^2
   d. none of the above

6. What is the most reasonable unit of measurement for a glass of orange juice?
   a. cm^2
   b. mL
   c. L
   d. g

7. A roll of pennies weighs about 125 grams. Monica saved until she had 15 rolls of pennies. About how much did her pennies weigh?
   a. 1,250 g
   b. 1,875 g
   c. 2,005 g
   d. 2,750 kg

8. What is the volume of the rectangular prism?
   a. 500 in.^{3}
   b. 600 in.^{3}
   c. 750 in.^{3}
   d. 15,000 in.^{3}

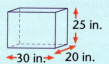

9. Find the perimeter of the triangle.
   a. 10.2 cm
   b. 11.8 cm
   c. 13.8 cm
   d. none of the above

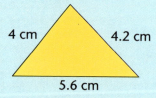

10. Mr. Simpson's yard is shaped like a rectangle. The length is 70 feet and the width is 50 feet. He spread grass seed on of the lawn. How many square feet were covered with grass seed?
    a. 875 ft^2
    b. 1,750 ft^2
    c. 2,000 ft^2
    d. 3,500 ft^2

CHAPTER 10

Statistics, Graphing, and Probability

Tania Harrison
Wellington, FL

10.1 Data and Statistics

Objective: to collect, organize, and analyze data

Each student in Mrs. Jenks's class completed a survey about his family. A survey asks questions to get information.

The third question on the survey was "How many brothers and sisters do you have?"

Factual information is often called **data**. When you study **statistics**, you collect, organize, and interpret data.

Mrs. Jenks collected the data by asking each student to write his answer to question 3 on the board. The **range** of the data is the difference between the greatest number and the least number in the set of numbers.

Brothers and Sisters						
3	3	0	2	1	0	2
2	1	3	1	2	4	2
0	2	1	2	3	2	4
1	3	1	1	2	3	0

The greatest number is 4 and the least is 0.

$$\text{range} = 4 - 0 \text{ or } 4$$

To organize the data, Mrs. Jenks drew a **frequency table**. It lists the data and tallies the time each number appeared in the class list.

▶ The frequency means how many times that number occurred.

Brothers and Sisters	Tally Marks	Frequency
0						4
1								7
2										9
3							6
4				2

TRY These

Use the frequency table above to answer questions 1–4.

1. How many students are in Mrs. Jenks's class?

2. What does the  mean in the tally mark column?

3. How many students in Mrs. Jenks's class have no brothers or sisters?

4. Which number of brothers and sisters received the most tally marks?

Find the range of each set of data.

1. 10, 20, 30, 40, 50
2. 84, 144, 89, 99, 93, 95
3. 2.5, 5, 7.5, 10, 12.5
4. 11.1, 40.3, 58.7, 16.6, 31.4
5. 163; 651; 136; 1,222; 566
6. 2,012; 2,012.2; 2,011.8; 2,012.6

Organize each set of data in a frequency table.

7.
Birth Month			
Nov.	Mar.	Apr.	Aug.
Jan.	Sept.	Dec.	Mar.
Nov.	May	Feb.	Dec.
Apr.	July	Oct.	Apr.

8.
Favorite Pet				
Dog	Dog	Cat	Fish	Dog
Hamster	Dog	Rabbit	Dog	Cat
Horse	Cat	Dog	Dog	Dog
Gerbil	Dog	Dog	Cat	Bird

Make a frequency table for the test score data. Then answer the questions.

9. What is the range of the data?
10. Which score appears most often?
11. Why is it easier to interpret these test scores in a frequency table rather than the list shown?

Test Scores							
95	85	90	85	75	70	80	85
90	80	80	75	80	70	80	95
80	75	100	90	90	65	95	85
80	75	60	75	85	80	85	80
90	80	100	85	80	80	85	90

MIXED Review

12. A survey of 300 pet owners found that $\frac{1}{3}$ own dogs, $\frac{1}{3}$ own cats, $\frac{1}{4}$ own both dogs and cats, and $\frac{1}{12}$ owns some other kind of pet. Find how many people fall in each group.

13. What is the greatest whole number that, when rounded to the nearest ten or to the nearest hundred, rounds to 2,500?

14. Two angles of a triangle have measures of 60° and 30°. What kind of angle is the third angle?

15. Mrs. Black buys 3 pounds of potatoes for $2.07. How much does she pay for each pound?

10.2 Mean, Median, and Mode

Objective: to find the mean, median, and mode of a set of data

The Department of Motor Vehicles is trying to find how many driver's licenses they issue on an average day. The number of licenses issued each day for 9 days is listed on the truck.

The **median** is one measure of organized data. To find the median, first list the numbers in order. The middle number is the median. When there are two middle numbers, the *median* is the mean (average) of the two middle numbers.

4, 5, 5, 6, 6, 7, 7, 7, 9

The middle number is 6.

The median is 6.

The **mode** is another measure of data. To find the mode, make a frequency chart. The mode is the most frequently listed number.

number	4	5	6	7	9
frequency	1	2	2	3	1

The most frequent number is 7.

The mode is 7.

The **mean** is the most common measure of data. When there are two middle numbers, the *median* is the mean of the two middle numbers. The mean is also called the **average**. To find the mean, add the numbers in the set and divide by the number in the set.

4 + 5 + 5 + 6 + 6 + 7 + 7 + 7 + 9 = 56

56 ÷ 9 = 6.22222222

sum of data   how many numbers

The Department of Motor Vehicles issues about 6 licenses each day.

The mean must be rounded to the nearest whole number because you cannot issue part of a license. The rounded mean is 6.

Order each set of data from least to greatest. Tell how many numbers are in the set and find the sum of the set of data.

1. 106   8   92   36   18
2. 2   3   1   4   1   2   1   2   3   1
3. 393   482   492   473
4. 9   5   18   5   7   8   12   1   4

Exercises

Find the median and mode.

1. 4, 4, 4, 5, 8
2. 3, 7, 7, 15, 18
3. 11, 11, 8, 7, 8, 6, 11
4. 2, 13, 4, 5, 6, 13, 15
5. 6, 11, 3, 4, 6, 4, 4
★6. 10, 9, 18, 14, 8, 10

Find the mean.

7. 6, 10, 13, 11
8. 6, 12, 14, 8, 20
9. 81, 75, 78, 79, 82
★10. 5, 6, 7, 3, 6, 8, 9, 10, 12

Problem Solving

11. Mr. Smith lists the ages of his employees and says that there is no mode. How can this be?

12. Mike's test scores are 80, 85, 91, 89, and 95. What is his mean score?

13. The average attendance for 4 football games was 1,566. What was the total attendance for the 4 games?

Constructed Response

14. A grocery store surveyed its customers to find which pudding they liked best. Which measure of data do you think was used? Why?

15. The students in Mrs. Smith's class had these scores for their last test: 100, 82, 82, 85, 93, 85, 71, 94, 84, 82, 88. Help Mrs. Smith by finding the mean, median, and mode of the test scores. Compare the mean, median, and mode of these scores. What might Mrs. Smith conclude?

Mixed Review

16. Mrs. Lewis is having a party for 43 people. A cake serves 8 people. How many cakes should she order from the bakery?

Compute.

17. 2,732 ÷ 8
18. 6,578 + 9,677
19. $54.70 ÷ 5
20. 5 × 390
21. 46 + 75 + 71
22. 59 × 24

10.3 Problem-Solving Application: Using Statistics

Objective: to solve problems using mean, median, and mode

The hospital staff members log how many births there are at the hospital each day. Dr. Chin announced that the number of births each day was 12.

Did he use the **mode**, **median**, or **mean**?

Sunday	11
Monday	7
Tuesday	15
Wednesday	11
Thursday	14
Friday	18
Saturday	8

1. READ — You know how many births there were on each day of the week. You know how many days are in the week and you know the statistic given by Dr. Chin.

2. PLAN — You can look at the data to see if the mode or median is a possibility.

3. SOLVE — The mode is 11. He did not use the mode. The median is also 11. He did not use the median.

Find the mean with this equation.

$n = (11 + 7 + 15 + 11 + 14 + 18 + 8) \div 7$
$n = 84 \div 7$
$n = 12$

Dr. Chin used the mean.

4. CHECK — Three numbers of births are over 12 and four are below 12. No number is more than six away from 12. So 12 is a reasonable statistic for this set of numbers.

Tell which measure would be best for each situation: *range, mean, median,* or *mode*.

1. Bart's test scores are 95, 96, 88, 90, and 94.
2. Louise surveys people to find what kind of pet they like.
3. Mrs. Wilson separates her bills into the most expensive and least expensive.
4. ChemCo finds an average salary for $65,000, $16,000, $12,000, $6,000, and $6,000.

A cereal company claims to have an average of 18 raisins in each bowl of Raisin Flakes. Ten bowls of cereal are poured.

1. What is the mean?
2. What is the median?
3. What is the mode?
4. Which measure did the company use? Why?

The salaries of some county sheriffs are listed below.

5. What salary is the mode?
6. What salary is the median?
7. What salary is the mean?
8. If a sheriff wanted to ask for a raise, which measure would the sheriff use? Why?

$40,000	$50,000
$42,000	$52,000
$50,000	$60,000
$50,000	$76,000
$50,000	$100,000

Use the chart to answer questions 9–10.

9. Leroy's Shoe Store kept a record of how many men's shoes were sold in each size. What measure is most useful when ordering shoes? Why?
10. How many pairs of shoes did Leroy's Shoe Store sell during this period?

Size	Sold	Size	Sold
$6\frac{1}{2}$	2	9	28
7	3	$9\frac{1}{2}$	42
$7\frac{1}{2}$	6	10	10
8	11	$10\frac{1}{2}$	6
$8\frac{1}{2}$	25	11	3

10.3 Problem-Solving Application: Using Statistics

10.4 Stem–and–Leaf Plots

Objective: to create and interpret a stem-and-leaf plot to show the distribution of data

Mrs. Shields' students were practicing for their physical agility tests. One portion of the test is the one-minute sit-ups. The students' practice results are listed below.

One-Minute Sit-Ups Results

20	33	48	49	39	62	28	35
44	51	55	28	49	50	39	43

One way to organize the data is to use a **stem-and-leaf plot**. In a stem-and-leaf plot, the data is organized from least to greatest. The ones digits are the **leaves** and the tens digits are the **stems**. Follow the steps below to make a stem-and-leaf plot.

Step 1: Write a title.

Step 2: Write the stems from least to greatest. Each tens digit is a stem.

Step 3: Write the leaves in order from least to greatest. The leaves are the ones digits. Write them next to the stems that match their tens digits.

Step 4: Write a key that explains how to read the plot.

One-Minute Sit-Ups Results

Stem	Leaf
2	0 8 8
3	3 5 9 9
4	3 4 8 9 9
5	0 1 5
6	2

Key: 3 | 9 = 48

Use the stem-and-leaf plot to answer questions 1–4.

1. What does 3 | 9 mean in the stem-and-leaf plot?
2. How many students were represented in the stem-and-leaf plot?
3. Find the median of the data.
4. In which group did the students have the most scores: twenties, thirties, forties, fifties, or sixties?

Exercises

Use the data to the right to answer questions 1–4.

1. Make a stem-and-leaf plot of the data.
2. What is the difference between the greatest score and the least score?
3. How many people are in the class?
4. Find the mean, median, mode, and range of the data.

Science Test Scores

75	83	89	92	100
97	79	84	83	88

PROBLEM Solving

Use the stem-and-leaf plot to the right for problems 5–8.

5. What does 4 | 3 mean in the stem-and-leaf plot?
6. What was the fastest car speed?
7. Find the mean, median, mode, and range of the data.
8. Suppose that an additional car speed of 60 mph was added to the data. How would the stem-and-leaf plot change?

Car Speeds (mph)

Stem	Leaf
2	5 8
3	0 0 2 4 6 6
4	3 8

Key: 3 | 0 = 30

TEST Prep

9. If you made a stem-and-leaf plot for the following data, which stem would have *no* leaves?

    20, 42, 24, 28, 16, 28, 6

    a. 0   b. 1   c. 2   d. 3

10.5 Bar Graphs

Objective: to create and interpret single- and double-bar graphs

Chris surveyed his class to find out what kinds of movies his classmates like to watch. A **bar graph** can be used to show different sets of data. Chris made a single-bar graph of the data collected from the boys in his class.

Favorite Movie Genre		
Genre	Boys	Girls
Action	3	0
Comedy	7	5
Drama	0	4
Science Fiction	6	3
Animation	1	1

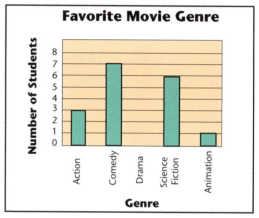

Chris wanted to compare the responses of boys and girls in a **double-bar graph**. Double-bar graphs are used to compare and contrast related information.

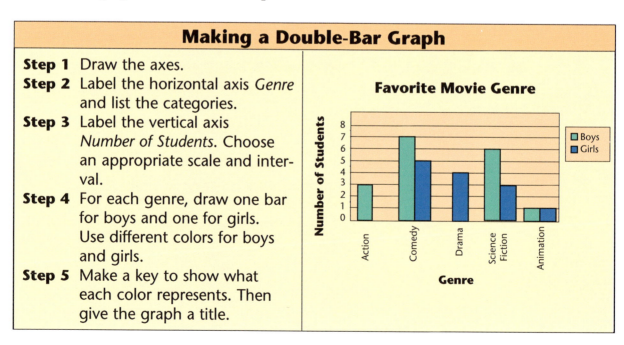

Use the double-bar graph to answer questions 1–3.

1. How many more boys like science fiction movies than girls?
2. For which genre is there the greatest difference between boys and girls?
3. Which movie genre is the same for boys and girls?

Use the data in the table to answer questions 1–5.

Mrs. Hall's fifth grade class sold bags of popcorn and bags of peanuts to raise money for a class trip. They want to compare the number of bags of each item sold.

Popcorn and Peanut Sale					
	Day 1	Day 2	Day 3	Day 4	Day 5
Popcorn	250	175	225	225	200
Peanuts	225	200	150	300	250

1. Make a double-bar graph showing the data from the table above.
2. How many more bags of popcorn were sold than peanuts on day 3?
3. On which day were the most bags of popcorn and peanuts sold?
4. For which day is the greatest difference between bags of popcorn and peanuts?
5. Was the average number of bags of peanuts sold each day more or less than 200?

Jacob, Emma, Matt, and Lily each tossed a penny 20 times. The results of how many times the penny landed on heads or tails are shown in the graph.

6. How many heads did Emma toss?
7. Which person tossed more heads than tails?
8. Which person had the least difference between the number of heads and tails tossed?
9. Find the total number of heads and tails tossed by all four people.

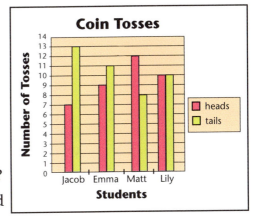

10.5 Bar Graphs

10.6 Line Graphs

Objective: to create and interpret single- and double-line graphs

Neil collected data for his meteorology project in science class. He made a **line graph** to display the data. A line graph is a way to organize data that shows change.

Day	Temperature (°F)
1	80
2	81
3	82
4	87
5	87
6	93
7	94

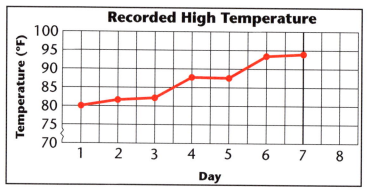

To compare two quantities that are increasing or decreasing over time, a **double-line graph** works better than a single-line graph.

Day	High	Low
1	80	67
2	81	66
3	82	69
4	87	66
5	87	69
6	93	76
7	94	77

To make a double-line graph:

Step 1: Draw the axes. Label the horizontal axis *Day* and the vertical axis *Temperature (°F)*.

Step 2: Plot the data for the high temperatures and connect the points. Use a different color to plot the data for the low temperatures and connect the points.

Step 3: Make a key to show what each color represents. Give the graph a title.

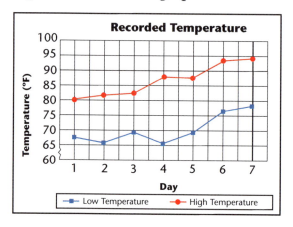

Use the double-line graph to answer questions 1–3.

1. What does the graph tell you about the high and low temperatures throughout the week?

2. For what day was the low temperature the greatest? the least?

3. For what day was there the greatest difference in the high and low temperature?

Use the data in the table to answer questions 1–4.

Number of Snowboards Sold				
	December	January	February	March
2005	112	90	84	76
2006	113	95	77	76

1. Make a double-line graph showing the data from the table above.

2. How many more snowboards were sold in 2006 during the month of January than 2005?

3. What conclusions can you draw from this graph?

4. Between what months of what year was there the greatest decrease in the number of snowboards sold?

The graph to the right shows the amount of money one homeroom class raised over a period of 5 weeks.

5. Which two weeks had the same amount of sales in candy bars? in magazines?

6. How much money has the class raised in candy bars? in magazines? both?

7. Between what weeks was there the greatest decrease in magazine sales? Why do you think this may have happened?

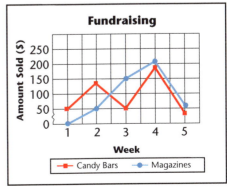

10.6 Line Graphs   **295**

10.7 Histograms

Objective: to create and interpret a histogram

The New York City Marathon is an event that draws racers from all different age groups. Look at this sample list of the ages of the participants for one year's race.

18	49	47	33	30	52	29	38	20	37
32	27	18	56	19	35	42	32	18	36
50	44	44	38	52	50	27	27	22	24
49	22	30	32	45	18	40	28	35	19

How did the number of participants in the 20–29 age group compare with that in the 50–59 age group? You can use a **histogram** to display and compare the data. A histogram is a graph that uses bars to display how frequently data occurs within an interval. You can make a histogram by following the steps below.

Intervals	Tally Marks	Frequency
10–19	𝍤 I	6
20–29	𝍤 IIII	9
30–39	𝍤 𝍤 II	12
40–49	𝍤 III	8
50–59	𝍤	5

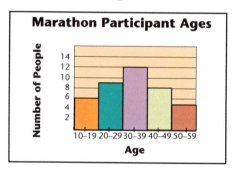

Step 1: Make a frequency table.

Step 2: Draw the axes. Label the horizontal axis and list the age intervals.

Step 3: Label the vertical axis and decide on a scale.

Step 4: Draw a bar for each interval.

Step 5: Give your graph a title.

Use the histogram to answer questions 1–3.

1. Which age group had the most participants?
2. How many more participants are in the 30–39 age group than in the 40–49 age group?
3. How many participants are younger than 20 years old?

Use the data below for problems 1–4. Use the intervals 0–4, 5–9, 10–14, and 15–19.

Broadway Shows Attended This Year						
2	6	10	4	5	6	0
1	3	8	3	4	15	11
5	9	12	0	8	9	7

1. Make a frequency table and a histogram for the data.
2. Find the mean, median, and mode of the set of data.
3. Which interval has the highest frequency?
4. Which interval has the lowest frequency?

Use the histogram at the right for problems 5–6.

5. The number of athletes who have been playing sports for 0–3 years is twice as many as the number in what other interval?

★6. The people who were surveyed have been playing sports for a combined total of 180 years. What is the mean number of years these athletes have been playing sports?

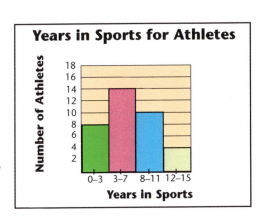

Constructed Response

7. When would you use a histogram instead of a bar graph to display data? Explain.

10.7 Histograms

10.8 Problem-Solving Application: Choose the Proper Graph

Objective: to choose the proper graph for a data set

The Chamber of Commerce of Parkville wants to make a graph to show how the city's population changed from 1980 to 2005.

They want to study how the population changed from one period to the next. Which type of graph should they use?

1980	87,500
1985	87,600
1990	86,000
1995	88,000
2000	87,000
2005	88,400

1. READ You know what years are being studied and the population in each of these years.

You need to know the best graph to use.

2. PLAN A circle graph is used to compare parts of a whole. A bar graph is used to compare given quantities or the frequency of items.

A line graph is used to compare quantities that are constantly changing.

They could use either a bar graph or a line graph. Since they want to study how the population changed they would probably use a line graph.

3. SOLVE Make a line graph. Let the horizontal scale be the years. The vertical scale is the population.

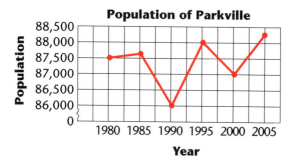

4. CHECK The line graph shows how the population increased and decreased. A line graph was a good choice for this set of data.

1. Which type of graph would you make from a frequency table?

2. Which type of graph would you make to show the temperature each hour of the day?

1. The Chamber of Commerce wants to show how many people attend the local football game each week. What graph should be used? Why?

2. The TV weatherman wants to show how the temperature in a certain city goes up and down each day. What kind of a graph should he use?

3. Mrs. Kent wants to show the scores her students received on the last math quiz in intervals. What graph should she use to show this? Why?

4. The fifth grade class took a poll to find out about their favorite foods. Pizza, hamburgers, and tacos were all favorites. What type of graph should the students use to show their findings?

5. Before Tyler leaves for a week's vacation, he discovers a leak in a water pipe. Putting a bucket under the leak, Tyler notices the bucket has $\frac{1}{2}$ inch of water after 12 hours, 1 inch after 24 hours, and 2 inches after 2 days. What type of a graph might Tyler use to show how fast the pipe was leaking?

6. The bicycling club wants to show how many miles each person biked last week using the tens and ones digits. What graph should they use to show their findings? Why?

MID-CHAPTER Review

Use the frequency table at the right.

1. What is the range of the scores?
2. Find the mode.
3. Find the median.
4. Find the mean.
5. Make a bar graph to show the results of the test.

Scores on a Math Test		
Score	Tally Marks	Frequency
60					3
90							5
95				2

10.8 Problem-Solving Application: Choose the Proper Graph

Problem Solving

Cross That Bridge When You Come to It

The city planners of Parkview built this fitness trail with seven bridges over the river. It was planned so that people could jog around the trail crossing each of the bridges only once. Did they succeed? (*Hint:* Start at one **X** and end at the other **X**.)

Extension

Where could the city planners build another bridge so that the plan works?

Cumulative Review

Compare using <, >, or =.

1. 987 ● 1,002
2. 3,628 ● 3,682
3. 0.2 ● 0.3
4. 0.4 ● 0.40
5. 2.3 ● 2.03
6. 15.6 ● 3.78
7. $\frac{2}{3}$ ● $\frac{3}{4}$
8. $\frac{6}{8}$ ● $\frac{9}{12}$
9. $\frac{9}{5}$ ● $\frac{9}{8}$

Compute.

10. 2,397 + 631
11. 7,816 − 4,307
12. 2.3 + 16
13. 6.2 − 3.19
14. 8 − 6.4
15. 605 × 7
16. 2.9 × 3.8
17. 972 ÷ 18
18. 2.5 ÷ 5
19. $\frac{3}{8} + \frac{1}{4}$
20. $\frac{5}{6} - \frac{1}{2}$
21. $1\frac{1}{2} - \frac{7}{8}$

Describe each angle as *acute*, *obtuse*, or *right*.

22.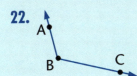
23.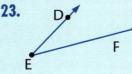
24.
25.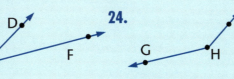

Complete.

26. 400 cm = ■ mm
27. 9 ft = ■ yd
28. 48 L = ■ mL
29. 3 T = ■ lb
30. 12 qt = ■ gal
31. 480 g = ■ kg
32. 56 d = ■ wk
33. 40 oz = ■ lb
34. 3 h = ■ min
35. 5 mi = ■ ft
36. 72 in. = ■ yd
37. 432 m = ■ km

Solve.

38. The perimeter of a square is 36 inches. What is the measure of each side?

39. Find the volume of a box whose length is 7.2 cm, width is 8 cm, and height is 3.6 cm.

10.9 Graphing Ordered Pairs

Objective: to locate and use ordered pairs

Maps often use **ordered pairs** to show us the location of buildings or parks.

The location of City Hall is given by the number pair (3, 2).

- The first number tells horizontal (↔) distance from 0.
  across
- The second number tells vertical (↕) distance from 0.
  up and down

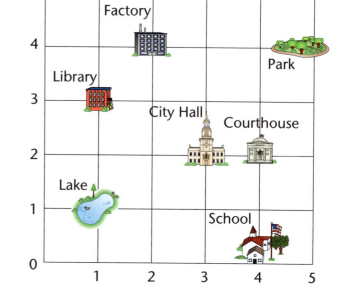

More Examples

A. What is located at (5, 4)?

Start at 0. Then go 5 units right and 4 units up.

The park is located at (5, 4).

B. Find the ordered pair that locates the school.

To get to the school from 0, you go 4 units right and 0 units up.

The school is located at (4, 0).

Use the grid to the right to name the letter for each ordered pair.

1. (2, 4)
2. (3, 3)
3. (10, 8)
4. (5, 0)
5. (6, 5)
6. (8, 10)
7. (0, 7)
8. (8, 2)
9. (2, 8)
10. (5, 6)

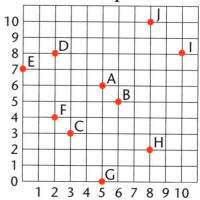

Use the grid to write an ordered pair for each point.

1. K
2. P
3. M
4. T
5. Q
6. L
7. S
8. I
9. J
10. N
11. U
12. R

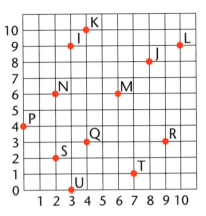

Draw a grid like the one above. Then graph and label each point.

13. V at (8, 8)
14. W at (4, 7)
15. X at (6, 1)
16. Y at (1, 6)
17. Z at (0, 3)
★18. O at (0, 0)

19. Graph the points on grid paper and connect them in order with segments.
(3, 1) to (6, 1) to (8, 3) to (8, 6) to (6, 8) to (3, 8) to (1, 6) to (1, 3) to (3, 1)
Name the figure you drew.

MIND Builder

Sets

Name the letter(s) inside each figure.

1. in the triangle
2. *not* in the rectangle
3. in the rectangle but *not* in the circle
4. in the triangle and in the circle
5. in the triangle and in the rectangle but *not* in the circle

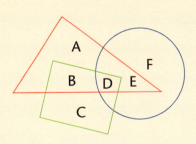

10.9 Graphing Ordered Pairs   303

10.10 Probability

Objective: to find the probability of events

When Arthur woke up, he heard there was a 75% chance of rain. He predicts that in 75 chances out of 100 it will rain.

Theoretical probability is the chance that an event will occur. When an event has a probability of 0, it is an **impossible event**. When an event has a probability of 1, it is a **certain event**.

The probability of an event can be expressed as $\frac{favorable\ outcomes}{possible\ outcomes}$.

More Examples

A. Suppose you toss a coin. What is the probability that it lands on heads?

Probability can be written as a fraction.

$\frac{1}{2}$ ← ways it can land heads up (favorable outcomes)
← possible ways to land (possible outcomes)

The probability of heads is $\frac{1}{2}$.

B. What is the probability the spinner will stop on yellow?

$\frac{2}{6}$ ← sections that are yellow
← possible sections

The probability of spinning yellow is $\frac{2}{6}$ or $\frac{1}{3}$.

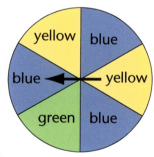

TRY These

Use the spinner above.

1. How many sections are blue?

2. What is the probability of spinning blue?

3. How many sections are green?

4. What is the probability of spinning green?

Solve. Use the jar of marbles.

1. Suppose you pick one marble without looking. What color are you most likely to pick?

2. What color are you least likely to pick?

3. What is the probability of picking a blue marble?

4. What is the probability of picking a red marble?

5. What is the probability of picking a blue or red marble?

6. What is the probability of picking a white marble?

7. What is the probability of picking a red or white marble?

8. What is the probability of picking a yellow marble?

9. What is the probability of *not* picking a green marble?

PROBLEM Solving

10. **Collect Data**
    Flip a foam cup 100 times. Record whether it lands on the bottom, top, or side. What is the probability that it will land on its bottom?

11. Suppose you roll a number cube that has the numbers 1–6 on it. What is the probability of rolling an even number?

★12. What is the probability of getting two heads if you flip a dime and a penny?

MIXED Review

13. A number has 3 as a factor if the sum of its digits has 3 as a factor. Find *four* numbers greater than 1,000 that have 3 as a factor.

Write each fraction in simplest form.

14. $\frac{9}{12}$ 15. $\frac{18}{24}$ 16. $\frac{20}{5}$ 17. $\frac{12}{8}$ 18. $\frac{36}{10}$

10.10 Probability   **305**

10.11 Experimental Probability

Objectives: to make predictions; to find the experimental probability of an event

When you perform an experiment, you may get different results than the predicted results from theoretical probability. **Experimental probability** is found by using the results from a game or an experiment.

Exploration Exercise

1. Place the marbles in a bag.
2. Pick one without looking. Record the color and return the marble to the bag.
3. Repeat this 23 more times.
4. Copy and complete the chart.
5. Compare your fractions to the answers to questions 3, 4, and 6 on p. 305.
6. Repeat the experiment. This time pick the marbles a total of 48 times.

Color	Number of Times	Fraction of Total
Red		$\dfrac{}{24}$
White		$\dfrac{}{24}$
Blue		$\dfrac{}{24}$

You may have found in the Exploration Exercise that your experimental probability is different from your theoretical probability. Theoretical probability is what is predicted will happen. Each time you perform an experiment, you may get different results.

The experimental probability of an event is the $\dfrac{\textit{favorable outcomes}}{\textit{number of trials}}$.

TRY These

The spinner below was spun. The results were recorded in the table. Use the table to find the experimental probability in questions 1–3.

Color	Tallies	Frequency
Blue	|||	3
Red	||||	5
Yellow	|||	3
Orange			1

1. red was spun
2. orange was spun
3. blue was spun

Exercises

Find the theoretical probability for rolling a number cube.

1. rolling a 3
2. rolling a 5
3. rolling a 4
4. rolling a 6
5. rolling a 2
6. rolling a 1

Copy the table below. Roll a number cube 20 times. Record the results in the table.

7. Find the experimental probability of rolling a 5.
8. Find the experimental probability of rolling a 3.
9. Compare the theoretical probability of rolling a 4 to the experimental probability of rolling a 4.

PROBLEM Solving

10. What is the theoretical probability of tossing a coin and getting tails? Complete 25 tosses and record your results. How many times did you toss tails?

11. How are theoretical and experimental probability alike? How are they different?

10.11 Experimental Probability

10.12 Counting Outcomes

Objective: to find possible outcomes using tree diagrams and multiplication

The town newspaper publishes the school lunch menu each day. Suppose you wanted to find how many different meals could be served that have a main course, beverage, and dessert.

Lunch Menu
Hamburger
Tuna Sandwich
Pizza
Milk
Juice
Apple
Brownie

Each combination is a possible outcome. You can use a **tree diagram** to show all the possible outcomes.

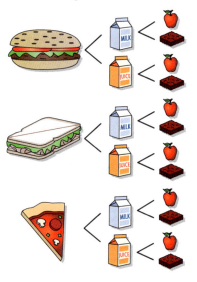

—hamburger, milk, apple
—hamburger, milk, brownie
—hamburger, juice, apple
—hamburger, juice, brownie
—tuna sandwich, milk, apple
—tuna sandwich, milk, brownie
—tuna sandwich, juice, apple
—tuna sandwich, juice, brownie
—pizza, milk, apple
—pizza, milk, brownie
—pizza, juice, apple
—pizza, juice, brownie

How many choices are there for a main course?
for a beverage?
for a dessert?

There are 3 × 2 × 2 or 12 possible outcomes.

What pattern do you notice in the tree diagram?

TRY These

Find the possible outcomes for tossing a dime and a nickel. Copy and complete the tree diagram.

___ 1. __heads__ _____

___ 2. _____ __tails__

___ 3. _____ _____

___ 4. _____ _____

Solve. Use a tree diagram. Use multiplication to confirm your answer.

1. The Snazzy Ice Cream Shoppe has chocolate, vanilla, and coffee ice cream. There is hot fudge and strawberry sauce. How many sundaes can be made using one flavor of ice cream and one sauce? Name them.

2. Enrico has blue, tan, and white jeans. He has brown, red, blue, and yellow shirts. How many different outfits are possible? Name them.

3. Mr. Green puts on a tie and a shirt for work each day. He has a solid red tie, a solid green tie, a striped tie, and a polka dot tie. His shirt colors are white, blue, green, and yellow. How many choices does Mr. Green have?

4. When Joe goes to the ball game, he has enough money to buy a hotdog and a drink. On his hotdog he can have catsup, mustard, or relish. He can choose iced tea or soda. How many combinations could Joe order?

5. Jacob's father plans to buy a new car. The dealer tells him red, black, and tan are his choices for interior colors. Red, white, green, blue, and black are choices for exterior colors. How many combinations of colors are possible?

★6. There are 5 colors and 6 shapes of balloons. How many combinations are possible if each balloon has only one color?

7. Tiffany wants to order soup and a sandwich. The soups are vegetable, chicken rice, and beef noodle. The sandwiches are tuna, turkey, ham, and veggie. How many choices does she have?

   a. 4   b. 8   c. 12   d. 16

10.12 Counting Outcomes

10.13 Using Statistics to Predict

Objective: to use statistics to predict

The mayor's office surveyed the people of the town to find out their opinion on a new tax issue. To save time they only called on those people who had even numbered house or apartment addresses.

Surveying part of a population is called getting a **sample**. Predicting how larger groups will answer the same questions is based on that sample.

Mayor Block asked if the people were in favor of the tax. Of the people surveyed, 4 out of 5 said yes. If he asks 35,000 people the same question, how many would say yes?

$\frac{1}{5}$ of 35,000 = 7,000 so $\frac{4}{5}$ of 35,000 = 28,000

THINK
$4 \times 7,000 = 28,000$

He would expect 28,000 of 35,000 to say yes.

Another Example

Mayor Block wants to know whether to spend city money to pave the road to the ballpark. He asks the Little League Parents Association for their opinion. Is this a good sample?

This is not a good sample. The Parents Association uses the road more than most people in the town. They would probably be in favor of paving the road. The rest of the town may not feel this is a good way to spend the money.

TRY These

Complete.

1. $\frac{1}{4}$ of 100 = 25, so $\frac{3}{4}$ of 100 = ■.

2. $\frac{1}{5}$ of 2,000 = ■, so $\frac{2}{5}$ of 2,000 = ■.

3. $\frac{1}{10}$ of 23,000 = ■, so $\frac{9}{10}$ of 23,000 = ■.

Complete.

1. $\frac{3}{5}$ of 500 = ■
2. $\frac{9}{10}$ of 3,000 = ■
3. $\frac{2}{3}$ of 9,333 = ■
4. $\frac{8}{9}$ of 81,000 = ■
5. $\frac{6}{7}$ of 42,777 = ■
6. $\frac{1}{2}$ of 312,000 = ■

Decide if each group is a good sample. Write *yes* or *no*. If it is not a good sample, tell why not.

7. asking 100 of 200 students about their favorite sport
8. asking 10 out of 4,000 people what television show they like
9. asking people in Arizona about the best brand of snow removal equipment
10. calling every tenth person in the phone book about how many times the phone is used each day

Problem Solving

11. A survey showed that 4 out of 5 people prefer yellow tennis balls. How many out of 5,000 people would prefer yellow tennis balls?
12. The probability of the mayor working on Saturdays is 1. How many Saturdays, out of 52, do you expect him to work?
13. Chin Lee knows the probability of rolling a 6 on a number cube. If he rolls the cube 60 times, how many times would he expect a 6?
14. Sarah exercises for 45 minutes on Monday, 30 minutes on Tuesday, 50 minutes on Wednesday, 25 minutes on Thursday, and 0 minutes on Friday. How many hours does she exercise?
★ 15. A floor tile measures 1 foot by 1 foot. How many tiles do you need to cover a floor that is 5 yards by 3 yards?

Constructed Response

16. Mr. Dennison buys a case of 12 rolls of film for $30. He later buys 3 rolls of film for $6. Which was the better buy? Explain.

Chapter 10 Review

LANGUAGE and CONCEPTS

Match.

1. the middle number in a set of data ordered from least to greatest
2. an average found by adding and dividing
3. sometimes used to find all possible outcomes
4. factual information used in statistics
5. how likely something will happen

a. data
b. mean
c. median
d. probability
e. tree diagram

SKILLS and PROBLEM SOLVING

Find the mean, median, mode, and range. (Sections 10.1–10.2)

6. 3, 4, 5, 4, 6, 6, 8, 4, 2

7. 68, 80, 70, 74, 81, 69, 70, 72

Use the stem-and-leaf plot to solve problems 8–9. (Section 10.4)

8. What were the highest and lowest scores?
9. How many students scored 88?

Math Test Scores	
Stem	Leaf
6	7 8
7	0 4 8 8 8
8	0 1 5 5 8 9 9

Use the double-bar graph to solve problems 10–12. (Section 10.5)

10. Which sport shows the greatest number of participants in fifth grade? in sixth grade?

11. By how much does the participation in basketball decrease from fifth grade to sixth grade?

12. How many sixth grade students participate in tennis?

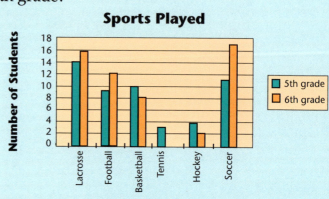

Draw and label a graph. (Sections 10.6–10.7)

13. Make a line graph.

Closing Price of M.A.T.H. Stock				
M	T	W	Th	F
$32	$28	$30	$32	$26

14. Make a frequency table and histogram.

Number of Candy Bars Sold					
25	32	33	41	55	63
36	42	35	61	52	23
26	37	38	42	27	39

Use the data to answer questions 15–17. (Sections 10.3, 10.8)

15. Find the best statistic for the attendance at the city council meetings.

16. What kind of graph would you use to show this data? Why?

17. Make a graph to show the data.

Attendance at the City Council Meetings	
January	215
March	200
May	225
July	260
August	200

A bag has 4 red gumballs, 6 orange gumballs, and 5 purple gumballs. Find the theoretical probability. (Section 10.10)

18. choosing an orange gumball

19. choosing a red or orange gumball

Find all possibilities. (Section 10.12)

20. There are large, medium, and small cups. There are tea, soda, milk, and water. How many different ways can a beverage be served?

Solve. (Section 10.13)

21. A survey found that 4 out of 5 people prefer popcorn over candy at the movies. How many people out of 2,000 prefer popcorn at the movies?

22. Why is asking 5 people in a school of 500 not a good sample?

Chapter 10 Review

Chapter 10 Test

For each set of data, find the range, mode, median, and mean.

1. 4, 6, 7, 10, 8, 3, 4
2. 20, 40, 10, 20, 50

Use the stem-and-leaf plot to answer questions 3–5.

3. What does 1 | 2 mean in this stem-and-leaf plot?
4. How many people are represented in the data?
5. How many people are 12 years old?

Ages of People at Summer Camp

Stem	Leaf
1	0 0 0 0 0 2 2 2 3 3 3 3
2	3 4 4
3	7
4	
5	2

Key: 2 | 4 = 24

6. Make a line graph.

High Temperatures (°F)

	M	T	W	Th	F
March	45°	50°	45°	55°	60°
July	85°	90°	85°	95°	80°

7. Make a histogram.

Candies in a Snack–Size Package

Number of Candies	Tally	Frequency
9–11	|||||		6
12–14	||||| ||||| |||	13
15–17	||||| ||||| ||||| ||||	19
18–20	|||||	5

Solve.

8. The science club has many books. Bryan counted 30 books on space, 65 books on physics, 25 books on biology, and 30 books on meteorology. What type of graph should he use? Why? Draw the graph.

9. What is the theoretical probability of tossing a coin and getting heads? Complete 20 tosses and record your results. How many times did you toss heads?

10. A survey says that 5 out of 6 people prefer gel toothpaste to regular toothpaste. If you asked 12,000 people which type of toothpaste they prefer, what number would you expect to say gel toothpaste?

Change of Pace

Means, Medians, and Magic Squares

A **magic square** is a puzzle in which you are asked to place nine numbers in a 3 × 3 array so that any horizontal, vertical, or diagonal sum is the same number. The square at the right is a magic square for the numbers 0–8. The sum in any direction is 12.

7	0	5
2	4	6
3	8	1

Have you ever wondered if there is a trick to quickly finding the correct arrangement?

Try these hints.

- Order the numbers from least to greatest. Which number is the median? Where is it located in the square?

- Now find the sum of all the numbers. What is the mean of these numbers? How does it compare to the median?

- Find the magic sum of a 3 × 3 square by dividing the sum of the numbers by 3.

- Find the mean of each row, each column, and each diagonal. How do they compare with the mean of the square?

What patterns do you see?

Copy each magic square. Use the hints to find the magic sum and total sum of each square. Then check to see if you are correct by completing the magic squares.

1.
8	1	
3		7
4	9	

2.
	2	12
6		14
8		4

3.
	3	18
9		21
	27	6

4.
17		
12		16
14	18	11

5.
32	4	24
12		
	36	8

Change of Pace 315

Cumulative Test

1.   5.02
    × 0.3
   a. 0.1506
   b. 0.156
   c. 1.506
   d. 15.06

2. Which fraction is equal to 0.75?
   a. $\frac{75}{10}$
   b. $\frac{75}{1,000}$
   c. $\frac{3}{4}$
   d. none of the above

3. Which fraction is equal to $\frac{8}{12}$?
   a. $\frac{2}{4}$
   b. $\frac{56}{84}$
   c. $\frac{75}{100}$
   d. $\frac{4}{3}$

4. 0.42 + 65.3 + 2.21 = ____
   a. 9.16
   b. 67.93
   c. 71.71
   d. 87.82

5. 7)5.327
   a. 0.761
   b. 7.61
   c. 76.1
   d. none of the above

6.   7.235
    − 0.718
   a. 6.507
   b. 6.517
   c. 6.523
   d. 7.953

7. What is the ordered pair that names the location of point C of the star?
   a. (3, 5)
   b. (5, 1)
   c. (5, 3)
   d. $(5\frac{1}{2}, 3\frac{1}{2})$

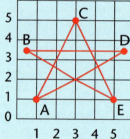

8. Reese ran 6 miles. Her time for each mile was 10.6 minutes, 8.2 minutes, 8.5 minutes, and 9.9 minutes. What is her average time for running a mile?
   a. 9.5 minutes
   b. 9.3 minutes
   c. 37.2 minutes
   d. none of the above

9. David bought 100 flower bulbs. On Saturday he planted 57 of the bulbs. Write the decimal that shows how many bulbs out of 100 he planted.
   a. 0.057
   b. 0.57
   c. 5.7
   d. 57.0

10. Eric pays $5.57 for 2.8 pounds of meat. How much does he pay per pound for the meat?

    What is a reasonable method for estimating the solution?
    a. $5.00 ÷ 2
    b. $5.57 ÷ 2.8
    c. $5.00 ÷ 3
    d. $6.00 ÷ 3

CHAPTER 11

Ratios and Percents

Heidi Vander Zouwen
Pewee Valley, KY

11.1 Ratios

Objective: to read, write, and simplify ratios

Samantha is selecting a book from the shelf. Three books are nonfiction and five books are fiction. One way to compare the number of nonfiction books to the number of fiction books is to write a **ratio**.

The numbers compared in a ratio are called **terms**. You can write a ratio as a fraction, with the first term as the numerator and the second term as the denominator.

Compare the books.

Step 1	Step 2
Identify the terms of the ratio. 3 nonfiction books 5 fiction books The first term is 3. The second term is 5.	Write the ratio of nonfiction books to fiction books. The ratio can be written in three ways: Word form: 3 to 5 Ratio form: 3:5 Fraction form: $\frac{3}{5}$ To read all three forms, say: 3 to 5.

The ratio of nonfiction books to fiction books is 3 to 5, 3:5, or $\frac{3}{5}$.

Examples

Express each ratio in simplest form.

A. 26 stickers to 4 sheets of paper

$$\frac{26 \div 2}{4 \div 2} = \frac{13}{2}$$, or 13:2, or 13 to 2

B. 12 ears of corn to 20 tomatoes

$$\frac{12 \div 4}{20 \div 4} = \frac{3}{5}$$, or 3:5, or 3 to 5

If both terms have a common factor, you can use that factor to simplify the ratio.

Write each ratio *three* different ways. Simplify if necessary.

1. 7 paints to 3 brushes
2. 15 beads to 5 sequins
3. 4 pens to 6 pencils
4. 12 stamps to 3 stamp pads
5. 8 dry erase markers to 10 dry erase boards
6. 32 books to 6 shelves

Write each ratio *three* different ways. Simplify if necessary.

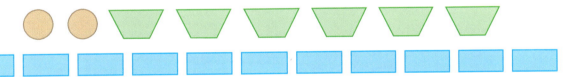

1. circles to rectangles
2. trapezoids to rectangles
3. circles to trapezoids
4. rectangles to other figures
5. circles to quadrilaterals
6. circles to all figures

PROBLEM Solving

7. Johan emptied a box of puzzle pieces onto the floor. He counted the number of each shape and organized his findings in the chart at the right. What is the ratio of end pieces to corner pieces?

Number of Each Type of Shape	
End pieces	20
Corner pieces	4
Middle pieces	25

8. Stacey has 35 yellow beads and 45 black beads to make a necklace. Find the ratio of the number of yellow beads to the number of black beads in simplest form.

9. Dan had $25 to spend at the fair. He spent $5 on food, $12 on rides and games, and saved the rest. Find the ratio of the amount of money spent on food to the amount saved.

Constructed Response

10. A rectangle measures 30 in. long by 50 in. wide. Find the ratio of the length to the perimeter of the rectangle. Explain your reasoning.

MIXED Review

Compute. Write each answer in simplest form.

11. $\dfrac{3}{8} + \dfrac{1}{2}$
12. $\dfrac{3}{5} + \dfrac{9}{10}$
13. $\dfrac{4}{9} - \dfrac{1}{3}$
14. $\dfrac{7}{10} - \dfrac{2}{5}$

15. $\dfrac{1}{9} \times \dfrac{3}{4}$
16. $\dfrac{7}{8} \times \dfrac{1}{3}$
17. $\dfrac{7}{8} \div \dfrac{3}{4}$
18. $\dfrac{4}{5} \div \dfrac{2}{5}$

Cumulative Review

Write the value of the underlined digit.

1. <u>2</u>,450,987.06
2. 38,<u>1</u>07,654
3. 56,789.0<u>5</u>4
4. 312,43.09<u>8</u>

Order from least to greatest.

5. 1.45   1.451   1.5   1.54
6. 3.1   3.01   3.11   3.011
7. 8.10   7.99   8.01   8.09
8. $4\frac{1}{4}$ $4\frac{3}{10}$ 4   4.03

Write the prime factorization for each number.

9. 8
10. 35
11. 37
12. 18

Multiply.

13. 82 × 40
14. 162 × 126
15. 488 × 315
16. 5 × 0.44
17. 48 × 1.35
18. 2.35 × 100
19. $8 \times \frac{2}{3}$
20. $\frac{2}{5} \times 2\frac{6}{7}$

Divide.

21. 52 ÷ 16
22. 255 ÷ 42
23. 779 ÷ 31
24. 0.21 ÷ 7
25. 3.6 ÷ 6
26. 7.28 ÷ 7
27. 5.6 ÷ 100
28. $\frac{2}{3} \div \frac{1}{3}$

Complete.

29. 10 c = _____ pt
30. 156 in. = _____ ft
31. 8 T = _____ lb
32. 3,520 yd = _____ mi

Solve.

33. Uniforms are made up of shirts in either of 2 styles; pants in blue, black, or brown; and 3 types of hats. How many different uniforms are there?

34. The hiking trail is $4\frac{1}{3}$ miles long. The campers hiked $\frac{2}{5}$ of the trail. How many miles did the campers hike?

11.2 Equivalent Ratios

Objective: to find equivalent ratios

Jessie wants to buy onions to make Mexican salsa. If she can buy 3 onions for 40¢, how many onions can she buy for 80¢? Ratios are similar to fractions. They are used to compare things.

You can use **equivalent ratios**.

onions → $\frac{3}{40} = \frac{\blacksquare}{80}$ ← onions
money → money

$\frac{3 \times 2}{40 \times 2} = \frac{6}{80}$

Find equivalent ratios like you find equivalent fractions.

Jessie can buy 6 onions for 80¢.

More Examples

A. Two tomatoes cost 75¢. How much do four tomatoes cost?

 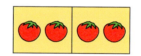

75¢ × 2 = 150¢
150¢ = $1.50

Four tomatoes cost $1.50.

B. For the salsa Jessie makes, the ratio of tomatoes to onions is 3 to 2. If she used 9 tomatoes, how many onions will she need? Jessie makes a ratio table.

Tomatoes	3	6	9	12	15
Onions	2	4	6	8	10

▶ Each number is a multiple of the first number in its row.

Jessie needs 6 onions.

TRY These

Replace each ■ so the ratios are equivalent.

1.

How many erasers for 90¢?

erasers → $\frac{1}{30} = \frac{\blacksquare}{90}$ ← erasers
cents → cents

2.

Two apples cost how much?

apples → $\frac{1}{20} = \frac{2}{\blacksquare}$ ← apples
cents → cents

Exercises

Copy and complete each ratio table. Then solve.

1. It takes 5 eggs to make 2 omelets.

Eggs	5	10	■	■
Omelets	2	4	6	8

How many omelets can you make with 20 eggs?

2. Tickets are 4 for $6.00.

Tickets	4	■	12	16
Dollars	6	12	■	■

How much will 20 tickets cost?

PROBLEM Solving

3. Jessie can serve 7 people with 2 bowls of salsa. How many bowls of salsa does she need to serve 28 people?

4. Jessie pays $2 for 6 tomatoes. How much would a dozen tomatoes cost?

5. If Jessie figures 2 quarts of iced tea for 8 guests, how many quarts will she need to serve 24 guests?

★6. Jessie gave 4 friends 5 tomatoes each. Then she gave 2 friends 3 tomatoes each. She had 6 left. How many tomatoes did she have at first?

MIND Builder

Equivalent Ratios

You can use lengths of shadows and equivalent ratios to find the height of tall objects such as flagpoles and buildings.
Compare the height of an object to the length of its shadow.

The flagpole is 20 feet high.

	person	flagpole
height of object	5	?
length of shadow	8	32

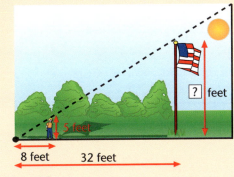

$$\frac{5 \times 4}{8 \times 4} = \frac{20}{32}$$

Use the ratio of 5:8 to find the height of objects with these shadows.

1. 56 feet   2. 24 feet   3. 72 feet   4. 112 feet

11.2 Equivalent Ratios   323

11.3 Rates

Objective: to use rates to solve problems

Barbara can type 100 words in 2 minutes. At that rate, how many words can Barbara type in 8 minutes?

A ratio that compares different units is called a **rate**. A **unit rate** is a rate in which the second term is one.

To find out how many words Barbara can type, find the unit rate and multiply.

Step 1	Step 2
Divide to find the rate in words per minute. *Per* means for each.	Multiply the unit rate by the number of minutes.
$\dfrac{100 \text{ words} \div 2}{2 \text{ minutes} \div 2} = \dfrac{50}{1}$	$\dfrac{50 \text{ words} \times 8}{1 \text{ minute} \times 8} = \dfrac{400 \text{ words}}{8 \text{ minutes}}$
She types 50 words per minute.	She can type 400 words in 8 minutes.

A rate shows words per unit of time as a speed. A slash (/) is often used for the word *per*. Fifty words per minute can be written as 50 words/min.

Example

Leah received $68 for 8 hours of baby-sitting. How much does Leah charge per hour?

$$\frac{68}{8} = 8\frac{1}{2}$$ Divide $68 by 8 hours to find the hourly rate.

The rate is $8.50 per hour or $8.50/hour.

324   11.3 Rates

Try These

Find the rate per unit of time.

1. 18 miles in 6 hours
2. $32 in 8 hours
3. 1,000 miles in 40 hours

Exercises

Find the rate per unit of time.

1. 128 miles in 16 min
2. 7.2 feet in 9 seconds
3. 49.5 meters in 3 minutes
4. 10.5 cm in 15 seconds
5. $125 in 5 hours
6. $73.50 in 7 hours

Find the distance traveled in the given amount of time.

7. 8 min at 12 m/min
8. 12 seconds at 14 ft/s
9. 5 hours at 75 mi/h
10. 0.5 hour at 40 mi/h
11. 0.2 second at 15 ft/s
12. 2.5 hours at 30 km/h

Find the length of time for each trip.

13. 360 mi at 30 mi/h
14. 660 ft at 15 ft/s
15. 495 km at 55 km/h

Problem Solving

16. The Calvert athletics department spent $786 on sneakers for boys' and girls' basketball teams. There are 12 players on each team. What was the unit price per pair of sneakers?

17. Morgan bought 3 Calvert sweatshirts for $76.50. What was the unit price per sweatshirt?

18. It took Rita 2 hours to go 35 miles during rush hour. At that rate, how long will it take to complete her 65-mile trip?

19. Mrs. Henry has been on the train for 3 hours and has traveled 240 miles. At this rate, her entire trip will take 6 hours. How far will she travel in all?

Constructed Response

20. Kenny can buy 3 pens for $2.70 or 5 pens for $4.00. Which is a better buy? Explain your answer.

11.3 Rates 325

11.4 Problem-Solving Strategy: Choose a Strategy

Objective: to choose a strategy to solve a problem

Aaron and his father biked 4.2 miles to the park. Then they biked on 3 trails that were each 2.5 miles long. After that, they biked home. How many miles did Aaron and his father bike altogether?

1. READ

What is the question?
How many miles did Aaron and his father bike altogether?

What do you know?
The distance to the park is 4.2 miles. They biked on three 2.5 mile paths.

2. PLAN

How can you find the answer?
You can make a drawing or you can write an equation.

3. SOLVE

Making a Drawing	Write an Equation
(drawing of house with trails: 4.2 mi, 2.5 mi, 2.5 mi, 2.5 mi) $4.2 + 2.5 + 2.5 + 2.5 + 4.2 = 15.9$	$(4.2 \times 2) + (2.5 \times 3) = n$ $8.4 + 7.5 = n$ $15.9 = n$

Aaron and his father biked 15.9 miles.

4. CHECK

Both methods give you the same result.
Your answer makes sense.

Solve. State the strategy you used to solve the problem.

1. Zach collects action figures of comic book heroes. The store sells packages of 2 figures for $9.52 and packages of 5 figures for $24.15. Which package offers the better unit price?

2. Five friends are waiting in a line to buy movie tickets. Chuck is in front of Becky but behind Brandon. Becky is in front of Lauren. Will is first. In what order are the friends waiting?

Solve. State the strategy you used to solve the problem.

1. Courtney runs around the lake every afternoon. She can run 2 miles in 15 minutes. If she can continue to run at that pace, how long will it take her to run 6 miles around the lake?

2. A store sells two different sizes of a model airplane. The ratio of the large size to small size is 4:3. How long is the small airplane if the larger one is 24 in. long?

3. Keith used triangular blocks to make the pattern shown. If he continues the pattern, how many blocks will there be altogether by the tenth row? What fraction of the blocks will be facing point downward?

4. Sarah has a $0.75 coupon she can use for anything at the grocery store. If she buys 6 yogurts for $0.65 each and 3 apples for $0.32 each, how much will she have to pay after she uses the coupon?

MID-CHAPTER Review

Write each ratio *three* ways.

1. 5 stripes to 9 dots
2. 12 cars to 18 SUVs
3. 14 dogs to 6 cats

Find the rate per unit of time.

4. 150 miles in 3 hours
5. 8,400 feet in 14 seconds
6. 8.10 meters in 3 minutes

11.5 Scale Drawings

Objective: to use equal ratios to interpret scale drawings

A **scale drawing** is a map or model of a real object that is enlarged or reduced by a specific factor. A **scale** is a ratio that compares the measurements in the drawing to the measurements of the real object.

How tall is the actual car that is modeled in the scale drawing?

You can use the scale and equivalent ratios to find the height of the car.

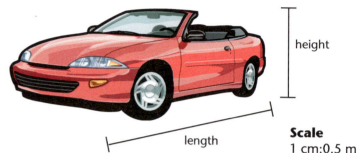

Scale
1 cm : 0.5 m

Step 1	Step 2
Use a centimeter ruler to measure the height of the car in the drawing. Measure to the nearest 0.5 centimeter. The car in the drawing is 2.5 cm tall.	Write the scale for the drawing in fraction form. $$\frac{1 \text{ cm}}{0.5 \text{ m}}$$
Step 3	**Step 4**
Write the scale and equivalent ratios to show the relationship between the height of the car in the drawing and the actual height of the car. $$\frac{1 \text{ cm}}{0.5 \text{ m}} = \frac{2.5 \text{ cm}}{h}$$	Find the height of the car. $$\frac{1 \text{ cm} \times 2.5}{0.5 \text{ m} \times 2.5} = \frac{2.5 \text{ cm}}{1.25 \text{ m}}$$ How tall is the car? The car is 1.25 m tall.

TRY These

1. Measure the length of the car in the drawing. Use this measurement to find the actual length of the car.

Exercises

In a scale drawing for a house, 1 cm represents 6 m. The scale is 1 cm:6 m. Find ■ in each case.

1. 4 cm represents ■ m.
2. ■ cm represents 12 m.
3. ■ cm represents 60 m.
4. 0.5 cm represents ■ m.
5. ■ cm represents 2.4 m.
6. 2.5 cm represents ■ m.

PROBLEM Solving

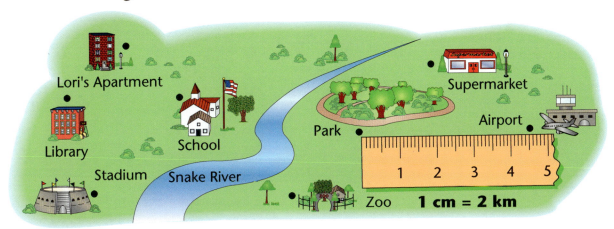

Solve. Measure to the nearest 0.5 cm.

7. On the map, what is the distance between Lori's apartment and the school?

8. What is the actual distance between Lori's apartment and the school?

9. On the map, what is the distance between Lori's apartment and the zoo?

10. What is the actual distance between Lori's apartment and the zoo?

11.5 Scale Drawings

11.6 Percents

Objective: to express ratios as percents

The word **percent** comes from Latin words that mean by the hundreds. Let the hundreds square at the right represent 1. Each small square is $\frac{1}{100}$, or 1 percent. One percent is written as 1%.

Fifty of the squares are colored. This means that $\frac{50}{100}$ or 50% is colored.

THINK Percent of a whole is like cents in a dollar.

What is another name for the colored part?

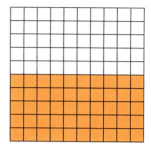

A penny is 1% or $\frac{1}{100}$ of a dollar.

Examples

A. Write the ratio $\frac{30}{100}$ as a percent.

$\frac{30}{100}$ written as a percent is 30%.

B. Write the ratio 45 out of 100 as a percent.

45 out of 100 written as a percent is 45%.

Write the percent of each grid that is colored.

1.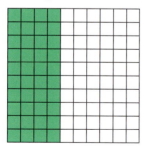

2.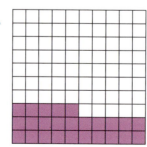

3.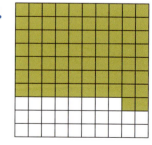

330   11.6 Percents

Exercises

Write each ratio as a percent.

1. $\frac{45}{100}$
2. $\frac{5}{100}$
3. $\frac{38}{100}$
4. $\frac{14}{100}$
5. $\frac{9}{100}$
6. $\frac{80}{100}$
7. $\frac{12}{100}$
8. $\frac{35}{100}$

9. 33 out of 100
10. 16 out of 100
11. 27 out of 100
12. 65 out of 100
13. 0 out of 100
14. 100 out of 100

Problem Solving

15. A survey showed that 92 out of 100 fifth graders like pizza. What percent of fifth graders like pizza?

16. Blake answered 87 out of 100 questions correctly on a test. What percent did he get correct?

Use the bar graph to answer questions 17–21.

17. What percent voted for yellow?

18. What percent voted for pink?

19. Which color received 12% of the votes?

★20. What percent voted for either blue or purple?

★21. Which two colors received exactly 17% of the votes?

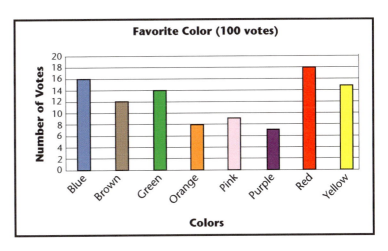

Favorite Color (100 votes)

11.6 Percents

11.7 Fractions, Decimals, and Percents

Objectives: to relate fractions, decimals, and percents; to write fractions, decimals, and percents in different forms

Approximately 65% of the human body is water. You can express 65% as a fraction or a decimal.

Write 32% as a fraction.	Write $\frac{1}{4}$ as a decimal.	Write 0.83 as a percent.
Write the number over 100. $$32\% = \frac{32}{100}$$ Simplify. $$\frac{32}{100} = \frac{8}{25}$$	Write the fraction as a division problem and divide. $$\frac{1}{4} = 1 \div 4 \quad \begin{array}{r} 0.25 \\ 4\overline{)1.00} \\ -8 \\ \hline 20 \\ -20 \\ \hline 0 \end{array}$$	Multiply by 100 by moving the decimal point two places to the right. Add a percent sign. $$0.83 \times 100 = 0.83$$ $$0.83 = 83\%$$

More Examples

A. Express 65% as a fraction in simplest form.

$65\% = \frac{65}{100}$ Write 65 over 100.

$65\% = \frac{13}{20}$ Simplify.

B. Express 65% as a decimal.

$65\% = 0.65$

$65\% = 0.65$ Divide by 100 by moving the decimal point two places to the left. Remove the percent sign.

 **TRY These**

Write each percent in decimal form and each decimal in percent form.

1. 0.66
2. 35%
3. 0.2

Express each fraction as an equivalent percent and each percent as a fraction in simplest form.

4. $\frac{3}{5}$
5. 40%
6. $\frac{2}{25}$

Exercises

Write each percent as a fraction in simplest form.

1. 50%
2. 42%
3. 80%
4. 95%
5. 70%
6. 25%
7. 8%
8. 125%

Write each fraction as a decimal.

9. $\frac{4}{25}$
10. $\frac{3}{4}$
11. $\frac{20}{50}$
12. $\frac{9}{10}$
13. $\frac{2}{5}$
14. $\frac{3}{25}$
15. $\frac{24}{25}$
16. $\frac{3}{20}$

Write each decimal as a percent.

17. 0.56
18. 0.85
19. 0.88
20. 1.75

Problem Solving

21. Casey has a total of 25 photographs for her photo album. She has 12 remaining photos to place. What percent of all her photos has she already placed?

22. Ty took a survey and found that $\frac{1}{5}$ of the students in his class wanted to take a field trip to a park. What percent of the students want to take a trip to the park?

23. There are 20 students in math class and 14 of them are girls. What percent of the students are boys?

24. The Calvert boys' soccer team won 4 out of 8 games. What percent of their games did they win?

Test Prep

25. What is 75% expressed as a decimal?

    a. 7.5
    b. 3.4
    c. 0.75
    d. 0.075

26. What is $\frac{4}{5}$ expressed as a decimal?

    a. 0.8
    b. 0.45
    c. 8.0
    d. 4.5

11.7 Fractions, Decimals, and Percents

11.8 Mental Math: Finding 10% of a Number

Objective: to use mental math to find 10% and multiples of 10% of a number

Bethany spends 20 hours per month training for a triathlon. If she spends 10% of the time running, how many hours per month does she spend running?

Compute. 10% of 20 = ■

There are two different methods to find 10% of a number.

Method 1	Method 2
Multiply by $\frac{1}{10}$. $10\% \times 20 = \frac{1}{10} \times 20$ $\frac{1}{10} \times \frac{20}{1} = \frac{20}{10}$ $= 2$	Move the decimal point to divide by 10. $20 \div 10 = 2.0$ $= 2.0$

Bethany spends 2 hours each month running.

Examples

A. Find 10% of 7.45.

$7.45 \div 10 = 0.745$

10% of 7.45 is 0.745.

B. Find 20% of 52.

$$\frac{1}{10} \times \frac{52}{1} = \frac{52}{10}$$

$$\frac{52}{10} = 5\frac{2}{10} = 5.20$$

THINK 20% = 2 × 10% $5.2 \times 2 = 10.4$

20% of 52 is 10.4.

TRY These

Find 10% of each number. Use mental math.

1. 65
2. 13
3. 9
4. 185
5. 3.9
6. 0.2

Exercises

Find 10% of each number. Use mental math.

1. 33
2. 1
3. 246
4. 100.6
5. 4.41
6. 7,939

Find 20% of each number.

7. 72
8. 315
9. 61
10. 48
11. 7,000
12. 1

PROBLEM Solving

13. Julianne says that 10% of 60 is greater than 20% of 30 because 60 is larger than 30. Is she correct? Explain your answer.

14. The bill for dinner was $58. Mary Ellen left a 20% tip. How much was the tip?

15. Justin went to the mall with $25.50. He spent 20% of his money on food. How much money was left?

Constructed Response

16. How could you use mental math to find 25% of a number?

11.8 Mental Math: Finding 10% of a Number

11.9 Percent of a Number

Objective: to find the percent of a number

Many families make a budget to help plan how they will use their money. The Conway family's weekly entertainment budget is shown to the right. Chris and Kate Conway would like to see a movie this week. It costs $9.25 to go to a movie theater. Does their budget allow them to go to the movie theater this week?

Total $80	
Dinner	50%
Movies	25%
Music	20%
Miscellaneous	5%

You can solve this problem by finding the percent of a number. The table below shows two different ways to find the percent of a number.

Method 1	Method 2
Write the percent in decimal form and multiply. $$25\% = 0.25$$ $$\begin{array}{r} 80 \\ \times\ 0.25 \\ \hline 400 \\ +160 \\ \hline 20.00 \end{array}$$	Write the percent as a fraction and multiply. $$25\% = \frac{25}{100} = \frac{1}{4}$$ $$\frac{1}{4} \times \frac{80}{1} = \frac{20}{1} = 20$$

Both methods confirm that Chris and Kate have $20 to spend on movies. Since the movie costs $9.25 per ticket, Chris and Kate would have to spend $18.50. According to their budget, they would be able to go to the movies this week.

Solve by writing the percent as a decimal.

1. 36% of 70
2. 22% of 18
3. 9% of 33
4. 80% of 77

Exercises

Solve by writing the percent as a fraction.

1. 60% of 40
2. 10% of 550
3. 50% of 76
4. 15% of 108
5. 15% of 20
6. 30% of 90
7. 25% of 40
8. 75% of 80

Solve. Use any method.

9. 76% of 100
10. 44% of 50
11. 8% of 111
12. 25% of 48
13. 8% of 82
14. 75% of 350
15. 40% of 520
16. 66% of 666

Problem Solving

17. Nick says his age is 20% of his father's age. His father is 45 years old. How old is Nick?

18. A magazine sells for $3.50. How much tax will be charged if the sales tax is 5% of the price?

★19. Donna shot 20 arrows during archery lessons. If 30% of the arrows hit the target, how many arrows did not hit the target?

Constructed Response

20. How could you use fractions to show 40% of 120?

21. Yasmine says that 15% of 20 is 30. Is Yasmine correct? Explain.

11.10 Circle Graphs

Objective: to create and interpret circle graphs

Mrs. Berrington's class was asked to create a **circle graph** with the amount of time the students spent on daily activities. Barrett made a table to show her percentages. Barrett then made a circle graph to visually show how she spent her day.

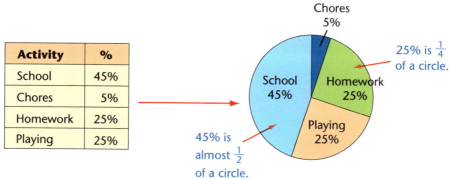

Examples

Forty students were surveyed. Use the circle graph at the right.

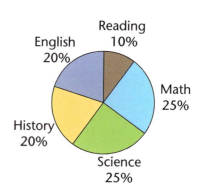

Favorite Subject

A. What percent of students chose English or math as their favorite subject?

    English   Math
    20%   + 25% = 45%   Add English and Math.

45% of students like English or Math.

B. How many students like science?

Find 25% of 40.   Convert 25% to a decimal.
0.25 × 40 = 10   Multiply by 40.

10 students like science.

338   11.10 Circle Graphs

Fifty-five students were surveyed. Use the circle graph at the right.

Favorite Ice Cream Flavor

1. What percent of students chose cookie dough or vanilla as their favorite ice cream flavor?

2. Which flavor of ice cream was the most popular?

3. Which flavor of ice cream was the least popular?

Use the circle graph at the right to answer problems 1–4.

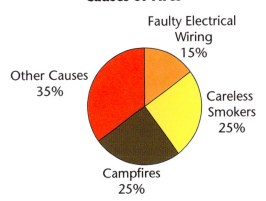

Causes of Fires

1. What fractional part of the total number of fires was the result of other causes?

2. What fractional part of the total number of fires was caused by careless smokers?

3. What fractional part of the total number of fires was caused by faulty electrical wiring?

4. What fractional part of the total number of fires was caused by campfires?

11.10 Circle Graphs   **339**

One hundred Calvert students were asked to tell their favorite type of book. The results of the survey are listed in the table below. Use the table to answer questions 5–8.

5. How many students chose literature as their favorite type of book?

6. How many students chose biographies or science as their favorite type of book?

7. What fractional part chose biographies? literature? science? the arts?

8. Trace the circle on plain paper. Then complete the circle graph by putting the category of books and the percentage of students in their proper places.

Books	Number of Students
Biographies	50
Literature	25
Science	20
The Arts	5

Problem Solving

Get a Strike

Vic's younger brother wants to try his hand at bowling, but he has set up the pins pointing in the wrong direction.

Help Fritz rearrange the pins to point in the opposite direction. Can you do it by moving only 3 pins?

Extension

If 15 pins were set up, what is the least number of pins you would move so that the pins point a different direction?

11.11 Problem-Solving Application: Discounts

Objective: to solve problems involving discounts

In order to make room for new releases, the video game store is having a sale and everything is advertised as 25% off. If a game sells for $32, what is its sale price?

1. READ — You need to find the sale price. You know the original price and the discount.

2. PLAN — Since the price is reduced by 25% you need to find the amount of discount and subtract it from the original price.

32 − (25% of 32) = ■

3. SOLVE

$32 − 8.00 = $24.00

4. CHECK — The sale price is $24. Add the amount of discount to the sale price to check.
$8 + $24 = $32

TRY These

Find the discount amount in dollars.

1. $12 book; 50% discount
2. $22 sweatshirt; 25% discount
3. $0.99 candy bar; 30% discount
4. $1.25 pen; 20% discount
5. $4.50 magazine; 10% discount
6. $0.48 pack of gum; 15% discount

Find the sale price.

1. $48 jacket; 33% discount
2. $12 shirt; 25% discount
3. $882 stereo; 17% discount
4. $150 MP3 player; 20% discount
5. $54 video game; 50% discount
6. $2,085 car; 60% discount

Solve.

7. 75% of $48
8. 15% of $490
9. 20% of $81
10. 50% of $1,750
11. 60% of $7.50
12. 30% of $54.90

PROBLEM Solving

13. The supermarket advertises a 20% off sale on steak. If 3 pounds of steak cost $20, what is the sale price for 3 pounds?

14. In July winter coats are put on sale for 25% off the regular price. What is the July price of a coat that normally sells for $104?

Constructed Response

15. Sara reads for 60% of an hour and Joan reads for $\frac{5}{12}$ of an hour. Who reads longer? Explain.

★16. The camera store is having a sale. Rick spent $275.00 for a camera. He also bought a case for $19.95 and a lens for $165.00. Before the sale those items would have totaled to $515.20. What was the discount? Explain.

17. Jennifer wants to buy a skirt that was originally $18. It is on sale for 40% off the original price. What is the sale price?

   a. $7.20   b. $25.20   c. $22.00   d. $10.80

Chapter 11 Review

LANGUAGE and CONCEPTS

Match.

1. the numbers that are compared in a ratio
2. a ratio that compares different units
3. means per hundred
4. a map or model of a real object that is enlarged or reduced by the same factor

a. scale drawing
b. percent
c. terms
d. rate

SKILLS and PROBLEM SOLVING

Write each ratio *three* different ways. Simplify if necessary. (Section 11.1)

5. 50 stars to 13 stripes
6. 4 quarters to 18 pennies
7. 14 students to 3 teachers

Copy and complete each ratio table. (Section 11.2)

8.
Cost	7		21
Tickets	2	4	

9.
Boys	5		15	
Girls	6	12		24

Find the rate per unit of time. (Section 11.3)

10. 200 miles in 5 hours
11. 3,000 feet in 6 seconds
12. 290 meters in 58 minutes

Write each ratio as a percent. (Section 11.6)

13. $\frac{52}{100}$
14. $\frac{99}{100}$
15. 27 out of 100

Write each percent as a fraction in simplest form. (Section 11.7)

16. 85% **17.** 30% **18.** 23%

Write each fraction as a decimal. (Section 11.7)

19. $\frac{4}{5}$ **20.** $\frac{6}{10}$ **21.** $\frac{3}{50}$

Write each decimal as a percent. (Section 11.7)

22. 0.32 **23.** 0.46 **24.** 0.4

Find 10% of each number. Use mental math. (Section 11.8)

25. 40 **26.** 4,500 **27.** 8.2

Solve. (Section 11.9)

28. Find 30% of 125. **29.** What is 75% of 85? **30.** What is 40% of 12?

Solve. (Sections 11.4–11.5, 11.10–11.11)
Use the graph below to answer questions 31–33. It represents a group of 75 students.

31. How many students liked comedy best?

32. How many students liked drama best?

33. How many students liked mysteries or biographies?

34. Kirk bought a sweater that was 20% off. If the original price was $35, what was the sale price?

35. In a scale drawing for an apartment 5 cm represent 1 m, so the scale is 5 cm:1 m. How many centimeters do 7 meters represent?

Favorite Movies

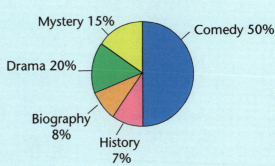

Chapter 11 Review

Chapter 11 Test

Write *two* equivalent ratios for each.

1. $\frac{1}{5}$
2. $\frac{3}{24}$
3. $\frac{5}{10}$
4. $\frac{3}{30}$

Complete each ratio table.

5.
Balls	2	4	6
Cents	95		

6.
Feet	1		5
Inches	12	36	

Write each fraction as a decimal and a percent.

7. $\frac{47}{100}$
8. $\frac{2}{5}$
9. $\frac{35}{50}$

Write each percent as a fraction in simplest form.

10. 20%
11. 35%
12. 38%

Solve.

13. 10% of 450
14. 25% of 400
15. 10% of 24.5
16. 20% of 90

A map is drawn with a scale of 1 cm:40 km. Find ■ in each.

17. 0.5 cm represents ■ km.
18. ■ cm represents 120 km.

Solve.

19. Kaylee is buying school supplies. A package of 5 pens costs $1.65. A pair of other pens costs $0.70. Which is the better price for 10 pens?

Sixty people were surveyed. Use the graph to answer problems 20–21.

20. What percent of people chose Indian or Japanese as their favorite type of food?

21. How many people chose Chinese as their favorite type of food?

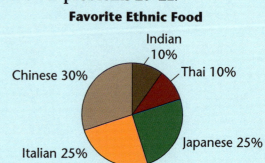

Favorite Ethnic Food

Indian 10%
Thai 10%
Japanese 25%
Italian 25%
Chinese 30%

Change of Pace

Scientific Notation

The Sun and Earth are part of the Milky Way Galaxy. But did you know that there are other galaxies in the universe? The Andromeda galaxy is 2.2 million light years away from Earth.

When you work with very large numbers, it can be difficult to keep track of the place value. You can write numbers like 2.2 million in scientific notation by using powers of ten.

2.2 million = 2.2 × 1,000,000
= 2.2×10^6

In scientific notation 2.2 million can be written as 2.2×10^6.

Examples

A. Write 475,000 in scientific notation.

4.75000   Move the decimal point five places to the left to get a number between 1 and 10.

The number of moves determines the exponent.

4.75×10^5 Count the number of original decimal places and rewrite the number using a power of 10.

The exponent determines the number of places to move.

B. Write 8.4×10^3 in standard form.

8.4   Move the decimal point three places to the right.

8,400   Add zeros if necessary.

Write each number in scientific notation.

1. 20   **2.** 680   **3.** 5,400   **4.** 7,000,000

Write each number in standard form.

5. 9.7×10^3 **6.** 8.88×10^4 **7.** 34.9×10^2

Cumulative Test

1. Which group shows all the numbers that are common factors of 24 and 36?
   a. 1, 2, 3, 4, 6, 12
   b. 1, 2, 3, 4, 6, 8, 12, 24
   c. 1, 2, 3, 4, 6, 9, 12, 18, 36
   d. 1, 2, 3, 4, 6, 8, 12

2. Which number is forty eight and thirty-four hundredths?
   a. 48.034
   b. 48.0034
   c. 48.34
   d. 48.34100

3. The art teacher has 826 bottles of paint. He wants to put the bottles into boxes that can hold 8 bottles each. How many boxes will he need?
   a. 103   b. 104
   c. 110   d. 113

4. A pattern of numbers is shown below.

   2, 3, 5, 8, 12, 17, ____

   If the pattern continues, what will be the next number?
   a. 22   b. 29
   c. 33   d. 23

5. Which group of figures contains only quadrilaterals?
   a.
   b.
   c.
   d.

6. What is the area of this rectangle?

   a. 54 sq ft   b. 27 sq ft
   c. 107 sq ft   d. 152 sq ft

7. Which of the following pairs shows a figure and its scale drawing?
   a.
   b.
   c.
   d.

8. A grocer posted the sign below near his fruit stand.

   Apples
   $0.50 each
   or
   $5.00 per dozen

   Tonya needs to buy 3 dozen apples to make pies. According to the sign, how much money will she save by buying apples by the dozen instead of individually?
   a. $3.00   b. $1.00
   c. $5.00   d. $2.00

9. Look at the figure below.

   Which statement about this figure is true?
   a. The figure has no parallel edges.
   b. The figure has exactly twelve vertices.
   c. The figure has exactly six faces.
   d. The figure has exactly eight edges.

10. Look at the spinner. Jack spins the arrow. What is the probability that the arrow will land on an even number?

    a. $\frac{4}{8}$ b. $\frac{5}{8}$
    c. $\frac{3}{8}$ d. $\frac{6}{8}$

348 Cumulative Test

CHAPTER 12

Looking Ahead

Megan DeBoard,
Honolulu, HI

12.1 Integers and the Number Line

Objectives: to identify integers on a number line; to compare integers

In a football game, a wide receiver caught the football and gained 5 yards. During the same game, a running back lost 3 yards. The numbers described above are examples of **integers**. Integers are positive and negative whole numbers.

5 yards gained: +5   3 yards lost: -3

To keep track of yards gained or lost, you can graph them on a number line. Extend the number line to the left and right of zero. A number line can also be used to compare integers. Numbers are greater in value as you move toward the right. Numbers are lesser in value as you move toward the left.

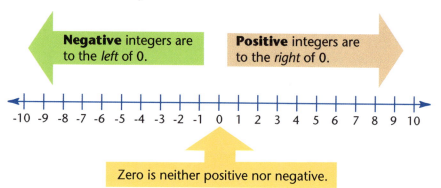

Negative integers are to the *left* of 0.

Positive integers are to the *right* of 0.

Zero is neither positive nor negative.

Examples

Compare using <, >, or =.

A. -4 ● 6

B. -3 ● -8

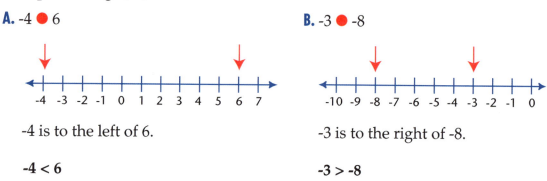

-4 is to the left of 6.

-3 is to the right of -8.

-4 < 6

-3 > -8

350 12.1 Integers and the Number Line

TRY These

Name the integer that is represented by each point.

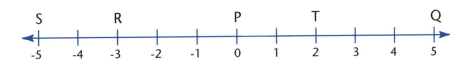

1. P
2. Q
3. R
4. S
5. T

Exercises

Write an integer to represent the situation.

1. a deposit of $100
2. 10° F below zero
3. a loss of 13 yards
4. 98 ft above sea level

Compare using <, >, or =.

5. -4 ● 2
6. -10 ● -3
7. 3 ● -3
8. 8 ● -9
9. 0 ● -2
10. -1 ● 4
11. -6 ● -5
12. -4 ● 7

PROBLEM Solving

13. Yuti went to the movies last night. She spent $10 for a move ticket. Write an integer for this situation.

14. The low temperature on Monday was -4° F. On Tuesday, the low temperature was 2° F. Was it colder on Monday or Tuesday?

15. The table shows the daily low temperatures for 5 days. Order the temperatures from least to greatest. On which day was it the coldest? the warmest?

★16. At 5:00 A.M. the temperature was -3° F. The temperature rose to 5° F by noon. How many degrees did the temperature rise?

Daily Low Temperatures	
Day	Low Temperature
Monday	1° F
Tuesday	-4° F
Wednesday	-7° F
Thursday	-2° F
Friday	0° F

12.1 Integers and the Number Line 351

12.2 Expressions and Variables

Objective: to write and evaluate expressions with variables

Damon bought 4 more baseball cards at the baseball card show on Saturday. He wants to know how many cards he now has in his collection.

He can write an **algebraic expression** and use a **variable** to represent the number of cards he already has. A variable is a letter or symbol used to represent some quantity.

Suppose Damon has 5 baseball cards.	Suppose Damon has 30 baseball cards.	Suppose Damon has n baseball cards.
Then he would have $5 + 4$ baseball cards.	Then he would have $30 + 4$ baseball cards.	Then he would have $n + 4$ baseball cards.

In the expression $n + 4$, n is the variable. The value of the expression changes depending on the value of n.

Translating Algebraic Expressions Key Words

+   add, more, plus, increased by
−   subtract, less, take away, decreased by

When you know the value of n, you can evaluate the expression. To evaluate an expression, replace the variable with the value and then compute.

Examples

A. Evaluate $m − 3$ if m is 12.
   $m − 3$ Replace m with 12.
   $12 − 3 = 9$ Subtract.

B. Evaluate $25 + p$ if p is 5.
   $25 + p$ Replace p with 5.
   $25 + 5 = 30$ Add.

Translate each phrase into an algebraic expression.

C. 7 more than a number
      $7 + n$

D. three subtracted from a number
      $n − 3$

Exercises

Evaluate each expression.

1. $3 + m$ if m is 6
2. $g - 4$ if g is 15
3. $9 + d$ if d is 5
4. $8 - e$ if e is 3
5. $h + 7$ if h is 20
6. $15 + t$ if t is 16
7. $17 - r$ if r is 9
8. $m + 5$ if m is 22

Match.

9. five more than a number
10. a number decreased by 12
11. 5 less than a number
12. $n + 7$

a. $n - 5$
b. $n + 5$
c. seven added to a number
d. $n - 12$

Write an algebraic expression for each phrase.

13. subtract 8 from a number
14. a number increased by 10
15. a number decreased by 2
16. add five to a number
17. take away ten from a number
18. three more than a number

Problem Solving

Use a variable to write an expression in problems 19–20.

19. Sara walked 5 dogs on Saturday. She also walked some dogs on Sunday. Write an expression to show how many dogs she walked altogether.

20. Josh got his weekly allowance. He used $2 to buy a hot dog. Write an expression to show how much money Josh has left from his allowance.

12.3 Mental Math: Solving Equations

Objective: to solve addition and subtraction equations using mental math

Joella had 6 charms. She received some additional charms for her birthday. She now has 10. How many charms did Joella receive for her birthday?

You can use an equation to help solve this problem. An equation shows that two expressions are equal. An example of an equation is shown below.

$3 + 5 = 8$ Both sides of the equation are equal to the same amount.

The equation that represents the situation above is $n + 6 = 10$, where n is the number of charms Joella received for her birthday. To solve this equation you can use mental math.

You need to think of a number that when substituted into the equation gives you 10.
What number plus 6 is equal to 10? The variable n is equal to 4, because $4 + 6 = 10$.
Joella received 4 charms for her birthday.

Examples

A. Solve. $5 - n = 3$

What number when subtracted from 5 gives you 3?

The number is 2, because $5 - 2 = 3$.

$n = 2$

B. Solve. $12 + n = 18$

What number when added to 12 gives you 18?

The number is 6, because $12 + 6 = 18$.

$n = 6$

TRY These

Are the expressions on both sides of the equals sign equal? Write *yes* or *no*.

1. $2 + 4 = 5 + 1$
2. $18 - 4 = 7 + 3$
3. $16 - 4 = 14$

Exercises

Solve using mental math.

1. $z + 3 = 6$
2. $b + 7 = 9$
3. $5 + h = 8$
4. $j + 1 = 8$
5. $d - 4 = 12$
6. $q - 4 = 8$
7. $g - 2 = 6$
8. $k - 2 = 10$
9. $y + 8 = 16$
10. $m + 2 = 14$
11. $4 - h = 4$
12. $b - 9 = 2$

PROBLEM Solving

Write an equation and solve.

13. In Mr. Carol's class, there are 16 students. If there are 9 boys, how many girls are in the class?

Constructed Response

14. Andy had 2 coupons with a combined value of $6. If 1 coupon was worth $2, how much was the other worth? Explain.

MIXED Review

Compute.

15. $\frac{3}{4} + \frac{2}{3}$
16. $\frac{1}{3} - \frac{1}{5}$
17. $\frac{5}{6} + 1\frac{1}{12}$
18. $4\frac{3}{4} - 2\frac{1}{2}$

19. $\frac{7}{8} \div \frac{3}{4}$
20. $\frac{6}{7} \times \frac{5}{8}$
21. $\frac{2}{5} \times 15$
22. $6\frac{1}{4} \times 1\frac{4}{5}$

12.3 Mental Math: Solving Equations

12.4 Order of Operations

Objective: to use the order of operations to simplify expressions

Betsy thinks that 5 × 3 + 2 is 17. Shani thinks 5 × 3 + 2 is 25. Who do you think is correct? Look at how both girls solved the problem.

Both girls computed parts of their problems correctly, but only one girl followed the correct order. Mathematicians have agreed to use the **order of operations** to ensure that an expression will have only one value.

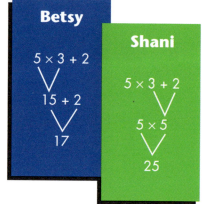

Order of Operations	Example
1. Complete all operations in grouping symbols (parentheses, brackets) first. 2. Multiply and divide in order from left to right. 3. Add and subtract in order from left to right.	(19 − 6) + 8 ÷ 2 = 13 + 8 ÷ 2 = 13 + 4 = 17

Betsy completed the problem correctly.

Use this mnemonic device to help you remember the order of operations.

Please	**E**xcuse	**M**y	**D**ear	**A**unt	**S**ally
Parentheses	**E**xponents	**M**ultiplication	**D**ivision	**A**ddition	**S**ubtraction

More Examples

Find the value.

A. 4 + 2 × 8 − 6

= 4 + 16 − 6   Multiply 2 and 8.

= 20 − 6   Add 4 and 16.

= 14   Subtract 6 from 20.

B. 9 − (16 − 13) + 15

= 9 − 3 + 15   Subtract 13 from 16.

= 6 + 15   Subtract 3 from 9.

= 21   Add 6 and 15.

TRY These

Which operation should be done first? Write *add*, *subtract*, *multiply*, or *divide*.

1. 2 × 9 + 6
2. 14 ÷ (5 + 2)
3. 12 − 8 ÷ 4 − 6
4. 10 − 7 + 2

Exercises

Find the value.

1. 20 − 8 × 2
2. (6 × 3) + (2 × 5)
3. 18 ÷ 2 + 9
4. 5 + 2 × 10
5. 16 + 4 × 3 − 28
6. 9 − 5 − 2 + 7
7. 10 + 8 × 2
8. (4 + 6) ÷ 2 + 3 × 5
9. 42 − (5 × 7) + 14

Compare using <, >, or =.

10. (54 ÷ 6) × 3 ● 54 ÷ (6 × 3)
11. 28 − (6 + 4) + 12 ● (28 − 6) + 4 + 12

PROBLEM Solving

12. Rewrite the following expression so that it equals 25. Add parentheses where needed.

    6 × 3 + 2 − 5

13. Look at the expression in problem 12. Simplify without parentheses.

12.4 Order of Operations

12.5 Problem-Solving Strategy: Working Backward

Objective: to solve problems by working backward

Ricky sold two times as many candy bars as Kemari. Kemari sold three times as many as Justin. Justin sold 7 candy bars. How many candy bars did Ricky sell?

1. READ — You need to know how many candy bars Ricky sold. You know that he sold twice as many as Kemari. You know that Kemari sold three times as many candy bars as Justin. You also know that Justin sold 7 candy bars.

2. PLAN — Here is a table to organize the information.

Ricky	Kemari's bars	× 2
Kemari	Justin's bars	× 3
Justin	7 candy bars	

3. SOLVE — Start with Justin. Work backward to find the number of candy bars Kemari sold: $7 \times 3 = 42$.

Kemari sold 21 candy bars.

Then use the number of candy bars that Kemari sold to find the number of candy bars that Ricky sold: $21 \times 2 = 42$.

Ricky sold 42 candy bars.

4. CHECK — Look back at your problem. Is your answer reasonable?

TRY These

Work backward to solve each problem.

1. You arrive at school at 8:15 A.M. It takes you 35 minutes to walk there. At what time should you leave for school?

2. Marco's apartment number is 31 greater than Anthony's. Anthony's apartment number is 43 less than Dominick's. Dominick lives in apartment number 105. What is Marco's apartment number?

Work backward to solve each problem.

1. Ten people got off a bus and three got on. If there are fifteen people on the bus now, how many were there at first?

2. Your lacrosse coach is retiring and you are saving money to buy her a gift. The gift will cost $30.00. If you save $4.25 per week for the next 3 weeks, you will have enough to buy the gift. How much did you already have?

3. You and Paige are planning a special banquet for the coach. The party will begin at 5:00 P.M. It will take you 30 minutes to set up the room. Before that, you have to spend $2\frac{1}{2}$ hours shopping for food and decorations. When will you leave to go shopping?

4. A store has 22 medium sweaters. There are 5 fewer large sweaters than medium ones. There are 6 fewer small sweaters than large ones. How many small sweaters are there?

MID-CHAPTER Review

Compare using <, >, or =.

1. 6 ● -1
2. -8 ● 0
3. -4 ● -2
4. -7 ● -8

Solve and check.

5. $t + 9 = 20$
6. $h - 6 = 14$
7. $7 - g = 4$
8. $c + 5 = 16$

Find the value.

9. $3 + 9 + 4$
10. $4 + (5 - 2) + 1$
11. $12 \div 4 + 2$
12. $3 \times (8 + 1) - 5$

12.5 Problem-Solving Strategy: Working Backward

Problem Solving

What a Deal!

Willie saved $85 to buy a bicycle. He bought a used bicycle for $40, sold it for $70, bought it back for $60, and then sold it again for $90.

Does Willie now have enough money to buy a better bike for $130?

Cumulative Review

Write the number as a percent and a decimal.

1. forty-seven hundredths
2. eight thousandths
3. twenty-five tenths
4. $\frac{1}{4}$
5. $\frac{1}{2}$

Find the missing angle.

6.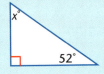

7.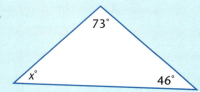

8. 23°, 68°, _____
9. 104°, 36°, _____

Multiply. Write the answer in simplest form.

10. $\frac{9}{5} \times \frac{1}{3}$
11. $\frac{12}{8} \times 1\frac{1}{2}$
12. $\frac{16}{4} \times \frac{2}{3}$
13. $\frac{21}{12} \times 1\frac{2}{3}$

Look at Lucy's schedule for the day and complete the table.

	Activity	Starting Time	Ending Time	Elapsed Time
14.	School	8:15 A.M.	3:05 P.M.	
15.	Piano lessons	3:45 P.M.		45 minutes
16.	Dinner		6:30 P.M.	35 minutes
17.	Homework	7:10 P.M.	8:03 P.M.	

Compare using <, >, or =.

18. 2.8 ● 2.80
19. 32.6 ● 3.28
20. 0.6 ● 0.52
21. $\frac{2}{3}$ ● $\frac{5}{6}$

12.6 Functions

Objective: to complete function tables

Mary walks 3 miles every day. If she walks for 1 day she walks 3 miles. If she walks for 2 days she walks 6 miles. How many miles does she walk if she walks for 4 days?

The relationship between the number of miles Mary walks and the days is called a **function**. A function is a rule that pairs each number of one set with a number from another set. For each *x* value there is only one *y* value.

The **input**, or number of days, is 4. The function is × 3. You can show this function with a function machine. The **output** is 12. So, Mary will walk 12 miles in 4 days.

You can also use a table to represent a function.

Example

Complete the table.

Rule: y = x + 5	
Input	Output
1	6
2	7
3	8
4	9

Substitute each input value into the function table to get an output value.

$y = 1 + 5 = 6$
$y = 2 + 5 = 7$
$y = 3 + 5 = 8$
$y = 4 + 5 = 9$

TRY These

Complete each function table.

1.

Rule: y = 3x	
Input	Output
1	3
2	■
3	9
4	■

2.

Rule: y = x − 2	
Input	Output
4	■
5	3
6	4
7	■

Exercises

Complete each function table.

1.
Rule: $y = 7x$	
Input	Output
0	
1	
2	
3	

2.
Rule: $y = x + 6$	
Input	Output
5	
7	
9	
11	

3.
Rule: $y = x - 1$	
Input	Output
12	
8	
4	
2	

4.
Rule: $y = 2x$	
Input	Output
0	
5	
8	
10	

5.
Rule: $y = 6x$	
Input	Output
4	
2	
8	
3	

6.
Rule: $y = 5 - x$	
Input	Output
2	
0	
4	
5	

PROBLEM Solving

7. Meredith's baby-sitting rate is described by the rule $y = 9x$, where y represents the amount of pay in dollars and x represents the number of hours she baby-sits. Make a table to find how much money Meredith earns if she baby-sits for 1 hour, 2 hours, 3 hours, and 4 hours.

8. The Marauders football team donates $25 to a local charity each time they score a touchdown. This can be represented by the function $y = 25x$, where y is the total amount of money donated and x is the number of touchdowns. Make a table and find the amount of money the team will donate if they score 1, 2, 3, and 4 touchdowns.

12.6 Functions 363

12.7 Graphing Functions

Objective: to graph functions

You can use the function machine for the function $x - 3$.

Remember, the function machine will subtract 3 from any number you put into it.

The table below shows the input and output for the function machine using the rule $x - 3$.

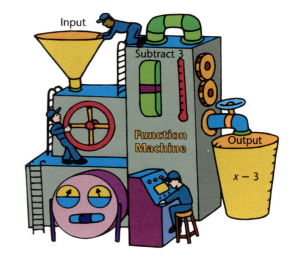

Rule: $y = x - 3$	
Input	Output
3	0
4	1
5	2
6	3

1. List the ordered pairs.
   (3, 0), (4, 1), (5, 2), (6, 3)

2. Graph the ordered pairs.

3. Connect the points with a straight line.

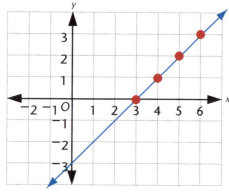

TRY These

Copy and complete the table for each rule. Then graph the ordered pairs on grid paper.

1.
Rule: $y = x + 4$	
Input	Output
0	
1	
2	

2.
Rule: $y = x - 2$	
Input	Output
4	■
5	3
6	4
7	■

3.
Rule: $y = 3x$	
Input	Output
1	
3	
5	

364   12.7 Graphing Functions

Exercises

Complete each function table. Then graph each function.

1.
Rule: $y = x + 3$	
Input	Output
0	
1	
2	
3	

2.
Rule: $y = x - 5$	
Input	Output
8	
7	
6	
5	

3.
Rule: $y = 4x$	
Input	Output
0	
1	
2	
3	

4.
Rule: $y = 8 - x$	
Input	Output
0	
1	
2	
3	

5.
Rule: $y = x + 3$	
Input	Output
0	
1	
2	
3	

6.
Rule: $y = x - 6$	
Input	Output
10	
9	
8	
7	

7.
Rule: $y = 4 - x$	
Input	Output
0	
1	
2	
3	

★8.
Rule: $y = 2x - 1$	
Input	Output
2	
3	
4	
5	

Problem Solving

9. Use the grid to the right to find the ordered pairs for each point. Then use the ordered pairs to find the function rule.

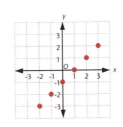

12.7 Graphing Functions

12.8 Problem-Solving Strategy: Write an Equation

Objective: to solve problems by writing an equation

A cellular phone company charges $0.05 for each minute spent talking and a $0.99 connection fee for each call. How much will you pay for a 20-minute phone call?

You can use equations to describe problem scenarios.

1. READ You need to find the cost for a 20-minute phone call. You know that you will be charged $0.05 per minute. You also know that you will be charged $0.99 for the connection fee.

2. PLAN You can write an equation to describe the problem scenario. Then you can solve the equation to find the total amount of money that you will pay.

3. SOLVE

Step 1	Step 2
Write an equation with words to show how the unknowns are related. charge per minute × number of minutes + connection fee = total $0.05m + $0.99 = t Let *m* be the number of minutes and *t* be the total.	Solve the equation. $0.05 × 20 + $0.99 = total = $1.99

You will pay $1.99 for a 20-minute phone call.

4. CHECK Did you correctly write and solve the equation?
Are your calculations correct?
Are your calculations reasonable?

Write an equation for each problem.

1. Pete is thinking of a number. He doubles it and then adds 15. The result is 29. What equation could you write to find his number?

2. A rectangle is 20 meters long and has a perimeter of 76 meters. Sheldon found the width of the rectangle without measuring. What equation did he use?

Write an equation for each problem. Then solve.

1. A triangle contains two marked angles and one unmarked angle. The two known angles are marked 60° and 42°. What is the measurement of the unmarked angle if the sum of the three angles is 180°?

2. Eight inches less than Eric's height is 50 inches. How tall is Eric?

3. Emily wants to order several pairs of capri pants that cost $15 each. The total shipping fee is $6. How much will Emily pay if she orders 4 pairs of capri pants?

4. Use problem 3 to answer the following question. How many pairs of capri pants can Emily order if she has $36 to spend?

Gears

Gear A turns clockwise. Gear B turns counterclockwise. Which direction does gear D turn?

Chapter 12 Review

LANGUAGE and CONCEPTS

Write the word that best completes the sentence.

1. _____ are positive and negative whole numbers.
2. A mathematical sentence that shows two expressions that are equal is called a(n) _____.
3. $7b$ means 7 _____ b.
4. The _____ is used by mathematicians to ensure that an expression will have only one value.
5. A rule that relates two variables is called a(n) _____.

> function
> times
> equation
> integers
> order of operations

SKILLS and PROBLEM SOLVING

Write an integer to represent the situation. (Section 12.1)

6. a withdrawal of $45
7. 1,600 feet above sea level
8. a gain of 15 yards
9. 7° F below zero

Compare using <, >, or =. (Section 12.1)

10. 6 ● -8
11. -1 ● 0
12. 8 ● -3
13. -4 ● -9

Translate each phrase into an expression. (Section 12.2)

14. a number added to sixteen
15. seven decreased by a number

Evaluate each expression. (Section 12.3)

16. $p + 7$ if p is 12
17. $18 - x$ if x is 6

Solve using mental math. (Section 12.3)

18. $h - 8 = 11$
19. $b + 12 = 24$
20. $v - 3 = 5$
21. $c + 9 = 19$

22. $7 + k = 12$
23. $m - 9 = 9$
24. $3 + z = 4$
25. $m - 2 = 8$

Find the value. (Section 12.4)

26. $4 + 16 \div 4 - 2$
27. $6 + (54 \div 6) \times 3$
28. $72 - (49 - 38) + 18$

Complete each function table. Then graph each function. (Sections 12.5–12.6)

29.
Rule: $y = 8x$	
Input	Output
1	
2	
3	
4	

30.
Rule: $y = x + 7$	
Input	Output
0	
1	
2	
3	

Solve. (Sections 12.4–12.5, 12.8)

31. Kory's lawn mowing rate is described by the rule $y = 15x$, where y represents the amount of pay in dollars and x represents the number of lawns that he has mowed. How much money does Kory earn for mowing 4 lawns?

32. When Ann Marie simplified the expression $3 \times 8 - 6 \times 4$, she said the result was 72. Do you agree with Ann Marie? Tell why or why not.

33. In Mrs. Peitz's class there are five more girls than boys. If there are 16 girls, how many boys are in the class?

34. Tommy has four times as many baseball cards as Greg. Greg has twice as many baseball cards as Leo. Leo has 45 baseball cards. How many baseball cards does Tommy have?

Chapter 12 Test

Write an integer to represent the situation.

1. 300 feet below sea level
2. a deposit of $150

Translate each phrase into an algebraic expression.

3. a number decreased by 9
4. seventeen added to a number

Compare using <, >, or =.

5. 4 ● 0
6. -3 ● 4
7. -8 ● 1
8. -6 ● -4

Solve using mental math.

9. $h - 4 = 16$
10. $b + 5 = 14$
11. $7 + m = 10$
12. $t - 8 = 10$

Find the value.

13. $(144 \div 12) + 5 \times 5$
14. $35 + (245 - 34) \times 2$
15. $(72 \div 8) + 4 \times 6$

Complete each function table.

16.
Rule: $y = 9x$	
Input	Output
0	
1	
2	
3	

17.
Rule: $y = x - 4$	
Input	Output
8	
7	
6	
5	

Solve.

18. Karina thinks that $6 + 7 \times 8$ is 104. Nicole thinks $6 + 7 \times 8$ is 62. Who do you think is correct? Explain.

19. Yvonne has twice as many pencils as Simone. Simone has four times as many pencils as Riley. If Riley has 4 pencils, how many pencils does Yvonne have?

20. There are three more cherry-flavored candies in a bag than orange-flavored candies. If there are 27 cherry-flavored candies in the bag, how many orange-flavored candies are in the bag?

21. The function $y = a + 28$ expresses Donna's age, a in terms of Sara's age, y. How old will Donna be when Sara is 45?

Change of Pace

Fibonacci Numbers

You can find the Fibonacci numbers in many aspects of nature such as in shells, plants, fruit, and vegetables.

The Fibonacci sequence is listed below. Can you figure out the pattern to find the next three terms?

$$1, 1, 2, 3, 5, 8, 13, \ldots$$

Follow the steps below to sketch the famous Fibonacci rectangles.

1. Draw two 1 cm × 1 cm squares side by side.

2. Draw a 2 cm × 2 cm square to sit on top of the two 1 cm × 1 cm squares.

3. Draw a 3 cm × 3 cm square to the right of the figure.

4. Draw a 5 cm × 5 cm square below the figure.

5. Continue drawing by adding larger squares around the previous figure in a clockwise direction.

Change of Pace

Cumulative Test

1. Melanie parks her bicycle in a parking lot and walks 300 yards to the entrance of the library. How many feet is 300 yards?
   a. 3,600 feet
   b. 100 feet
   c. 900 feet
   d. 600 feet

2. Which of the following numbers is less than 9.001?
   a. 9.011
   b. 9.0
   c. 9.101
   d. 9.1

3. Kenny can drive to the lake in 3 hours at an average speed of 45 miles per hour. Which equation can be used to find the distance Kenny drives to the lake?
   a. $3 + 45 = n$
   b. $45 \times 60 = n$
   c. $45 \times 3 \times 60 = n$
   d. $3 \times 45 = n$

4. $0.75 + 65.3 + 2.21 = $ _____
   a. 9.49
   b. 68.26
   c. 88.15
   d. 94.9

5. Find the area of the right triangle.
   a. 9 cm²
   b. 18 cm²
   c. 36 cm²
   d. none of the above

6. Samantha spent $34 on groceries. Her mom will pay her half of the cost of the groceries. Which expression should Samantha use to figure out how much money her mom will pay her?
   a. $2 + 34$
   b. $34 - 2$
   c. 34×2
   d. $34 \div 2$

7. A rectangular prism is shown below. What is the volume of the rectangular prism?
   a. 4 cm³
   b. 9 cm³
   c. 10 cm³
   d. 20 cm³

8. The table shows the ages of four teachers. Which teacher's age is a prime number?
   a. Tracy
   b. Will
   c. Kathryn
   d. Lawrence

Name	Age
Tracy	22
Will	40
Kathryn	53
Lawrence	39

9. Kendra ran 4 miles. Her times for each mile were 8.4 minutes, 7.2 minutes, 7.5 minutes, and 6.9 minutes. What is her average time for running a mile?
   a. 7.5 minutes
   b. 7.8 minutes
   c. 30.0 minutes
   d. none of the above

10. Which single transformation is represented by the two figures?

    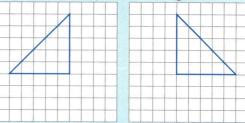

    a. rotation
    b. reflection
    c. translation
    d. none of the above

APPENDIX

Mathematical Symbols

$=$	is equal to
°	degrees
%	percent
$>$	is greater than
$<$	is less than
π	pi
$3a$	3 times a
4:3	ratio of 4 to 3
$\cong$	is congruent to

△ or △ sides are congruent

▭ or ▭ sides are parallel

└ right angle

$0.\overline{3}$ repeating decimal; $0.3 = 0.333...$

$\approx$ is approximately equal to

$\overleftrightarrow{AB}$ line AB

$\overline{AB}$ line segment AB

$\overrightarrow{AB}$ ray AB

△ triangle

∠ angle

⊥ is perpendicular to

∥ is parallel to

+4 positive four

-4 negative four

Formulas

$C = 2 \times \pi \times r$
circumference of a circle

$C = \pi \times d$
circumference of a circle

$A = \ell \times w$
area of a rectangle

$A = b \times h$
area of a parallelogram

$A = \frac{1}{2} \times b \times h$
area of a triangle

$P = 2 \times \ell + 2 \times w$
perimeter of a rectangle

$V = \ell \times w \times h$
volume of a rectangular prism

Metric System of Measurement

Prefixes
kilo (k) = thousand
hecto (h) = hundred
deka (da) = ten
deci (d) = tenth
centi (c) = hundredth
milli (m) = thousandth

Length
1 centimeter (cm) = 10 millimeters (mm)
1 meter (m) = 100 centimeters or 1,000 millimeters
1 kilometer (km) = 1,000 meters

Mass
1 gram (g) = 1,000 milligrams (mg)
1 kilogram (kg) = 1,000 grams
1 metric ton (t) = 1,000 kilograms

Capacity
1 liter (L) = 1,000 milliliters (mL)
1 kiloliter (kL) = 1,000 liters

Customary System of Measurement

Length
1 foot (ft) = 12 inches (in.)
1 yard (yd) = 3 feet or 36 inches
1 mile (mi) = 1,760 yards or 5,280 feet

Weight
1 pound (lb) = 16 ounces (oz)
1 ton (T) = 2,000 pounds

Capacity
1 cup (c) = 8 fluid ounces (fl oz)
1 pint (pt) = 2 cups
1 quart (qt) = 2 pints
1 gallon (gal) = 4 quarts

GLOSSARY

A

acute angle 8.2 An angle whose measure is between 0° and 90°.

acute triangle 8.3 A triangle that has three acute angles.

addend 2.1 A number that is added to another number. In 5 + 3 = 8, the addends are 5 and 3.

Addition Property of Equality 2.1 If you add a quantity to one side of an equal sign, you must add the same quantity to the other side to keep the two sides equal.

$$(3 + 2) + 5 = (2 + 3) + 5$$

algebraic expression 12.2 An expression involving variables, numbers, and operations.

angle 8.1 A figure formed by two rays that have a common endpoint.

area 9.3 The number of square units needed to cover a region.

Associative Property of Addition 2.1 Changing the grouping of three or more addends does *not* change the sum.

$$(3 + 2) + 5 = 3 + (2 + 5)$$

Associative Property of Multiplication 3.1 Changing the grouping of the factors does *not* change the product.

$$(3 \times 2) \times 5 = 3 \times (2 \times 5)$$

average 10.2 One number (or value) that best represents a set of numbers. Usually the *mean* is used as the average, but sometimes the *median* or the *mode* is the better measurement.

B

bar graph 10.5 A graph using bars to show a comparison between different values.

base of a geometric figure 9.6 The bottom side or face of a geometric figure.

C

capacity 1.9 The amount of liquid, grain, and so on, that a container can hold.

centimeter (cm) 1.9 A unit of length in the metric system.

$$1 \text{ centimeter} = 0.01 \text{ meter}$$

certain event 10.10 An event that has a probability of 1.

chord 8.7 A segment whose endpoints lie on a circle. The longest chord on a circle is the diameter.

circle 8.7 A curved figure in a plane. All the points of a circle are the same distance from a point (center) in the plane.

circle graph 11.10 A graph using a circle to compare parts of a whole.

circumference 9.2 The distance around a circle.

common denominator 6.3 The same denominator for two or more fractions. You can use 12 as a common denominator for $\frac{1}{3}$ and $\frac{3}{4}$ since $\frac{1}{3} = \frac{4}{12}$ and $\frac{3}{4} = \frac{9}{12}$.

common factor 5.4 A factor that two or more numbers have in common. The common factors of 8 and 12 are 1, 2, and 4.

Factors of 8: 1, 2, 4, 8
Factors of 12: 1, 2, 3, 4, 6, 12

common multiples 6.3 Multiples that two or more numbers have in common. Some common multiples of 2 and 3 are 6, 12, and 18.

Multiples of 2: 2, 4, 6, 8, 10, 12, 14, 16, 18, ...
Multiples of 3: 3, 6, 9, 12, 15, 18, 21, ...

Commutative Property of Addition 2.1 Changing the order of the addends does *not* change the sum.

$$3 + 2 = 5 \qquad 2 + 3 = 5$$

Commutative Property of Multiplication 3.1 Changing the order of the factors does *not* change the product.

$$3 \times 2 = 6 \qquad 2 \times 3 = 6$$

compass 8.7 A tool for constructing circles.

compatible numbers 4.1 Numbers that work well together.

composite number 5.2 A whole number that has more than two different factors. The number 4 is composite since it has three factors (1, 2, and 4).

cone 8.11 A solid that has a circular base and a surface from the boundary of the base to the vertex.

congruent figures 8.8 Figures that have the same size and shape.

corresponding parts of congruent figures 8.8 The parts (such as sides and angles) of congruent figures that match.

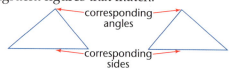

cube 8.11 A solid figure that has 6 square faces of equal sides.

cup (c) 6.11 A unit of capacity in the customary system of measurement.

customary system 6.11 A system of units of measure commonly used in the United States. The inch, foot, yard, and mile are units of length in the customary system. The cup, pint, quart, and gallon are units of capacity. The ounce, pound, and ton are units of weight.

cylinder 8.11 A solid with two circular faces that are congruent and a cylindrical surface connecting the two faces.

D

data 10.1 Factual information.

decimal 1.6 A fraction or mixed number that is written using place-value position.

$$\frac{4}{100} = 0.04 \qquad 3\frac{7}{10} = 3.7$$

decimeter (dm) 1.9 A metric unit of length.

$$1 \text{ decimeter} = 0.1 \text{ meter}$$

degree (°) 8.2 A unit used to measure angles. It is $\frac{1}{360}$ a circle.

denominator 5.5 In the fraction $\frac{3}{4}$, the denominator is the 4. It tells the number of equal-sized parts in the whole.

diameter 8.7 A line segment through the center of a circle with endpoints on the circle.

digit 1.1 The word used to name numbers in a place-value system.

Distributive Property 3.4 You can multiply a sum by multiplying each addend separately, and then adding their products.

$$6 \times (3 + 5) = 48$$
and
$$(6 \times 3) + (6 \times 5) = 48$$

dividend 4.1 A number that is divided by another number.

$$45 \div 9 = 5 \quad 9\overline{)45}$$

divisible 5.1 A whole number is divisible by another if, upon division, the remainder is zero. When you divide 12 by 3, the remainder is zero. So 12 is divisible by 3.

division 4.1 The operation used to separate the total number into equal groups, or to determine the number of equal groups. Division reverses multiplication.

divisor 4.1 A number by which another number is divided.

$$45 \div 9 = 5 \quad 9\overline{)45}$$

double-bar graph 10.5 A graph in which data are compared by means of pairs of rectangular bars drawn next to each other.

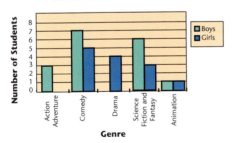

double-line graph 10.6 A graph that is used to show data by means of two broken lines.

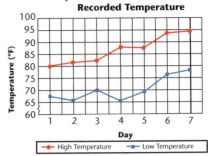

E

edge 8.11 A segment where two faces of a solid figure meet.

equation 12.3 A mathematical problem written horizontally using a letter to represent the unknown number.

equilateral triangle 8.3 A triangle that has three congruent sides.

equivalent fractions 5.5 Two or more fractions are equivalent if they name the same number.

$$\frac{1}{2} = \frac{2}{4} = \frac{3}{6}$$

equivalent ratios 11.2 Two ratios that are equal.

estimate 2.3 To find an approximate answer. An estimate of the sum of 484 plus 337 is 800.

expanded form 1.2 The form of a number that shows the value of each digit. 2,000 + 300 + 40 + 8 is the expanded form for 2,348.

experimental probability 10.11 Probability found when an experiment is performed.

Glossary 377

F

face 8.11 Flat surface of a solid figure.

factor 5.4 A number that is multiplied. Since 5 × 3 = 15, 5 and 3 are factors of 15.

factor tree 5.3 A method of finding the prime factorization of a number by using a diagram.

fluid ounce (fl oz) 6.11 A unit of capacity in the customary system of measurement.

foot (ft) 6.11 A unit of length in the customary system of measurement.

$$1 \text{ foot} = 12 \text{ inches}$$

fraction 5.5 A number for part of a whole or group, such as $\frac{1}{3}$.

frequency table 10.1 A way to organize data that lists the data, shows tally marks, and gives the number of times each piece of data appears.

function 12.6 A rule that gives exactly one value of y for every value of x.

G

gallon (gal) 6.11 A unit of capacity in the customary system of measurement.

$$1 \text{ gallon} = 4 \text{ quarts}$$

geometry 8.1 The study of lines and shapes.

gram (g) 1.9 A metric unit of weight.

$$1 \text{ gram} = 1{,}000 \text{ milligrams}$$

greatest common factor (GCF) 5.4 The greatest whole number that is a factor of two or more given numbers. The GCF of 12 and 18 is 6.

H

hexagon 8.4 A polygon with six sides.

histogram 10.7 A graph in which bars are used to display how frequently data occurs within equal intervals.

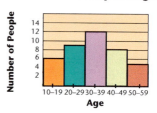

I

Identity Property of Addition 2.1 Any number added to zero has a sum equal to the original number.

$$5 + 0 = 5 \qquad 3 + 0 = 3$$

Identity Property of Multiplication 3.1 If you multiply any number by one, the product is that number.

$$5 \times 1 = 5 \qquad 1 \times 8 = 8$$

improper fraction 5.9 A fraction in which the numerator is greater than or equal to the denominator, such as $\frac{7}{4}$, $\frac{8}{8}$, or $\frac{9}{3}$.

impossible event 10.10 An event that has a probability of zero.

inch (in.) 6.11 A unit of length in the customary system of measurement.

input 12.6 The number you put into a function.

integers 12.1 Positive whole numbers and their opposites (negative numbers) and 0.

intersecting lines 8.1 Lines that meet or cross at a common point.

isosceles triangle 8.3 A triangle that has two congruent sides.

K

kilogram (kg) 1.9 A metric unit of weight.

1 kilogram = 1,000 grams

kilometer (km) 1.9 A metric unit of length.

1 kilometer = 1,000 meters

L

least common denominator (LCD) 6.3 The least common multiple of the denominators of two or more fractions.

least common multiple (LCM) 6.3 The least multiple, other than 0, common to sets of multiples. The LCM of 3 and 4 is 12.

leaves 10.4 The units digit written to the right of the vertical line in a stem-and-leaf plot.

line 8.1 A straight, continuous and unending set of points.

line graph 10.6 A graph in which line segments are used to compare changes in data.

line of symmetry 8.9 A line that separates a figure into two halves that match exactly.

line of symmetry

line segment 8.1 A part of a line that has two endpoints, such as $\overline{DE}$.

liter (L) 1.9 A metric unit of capacity.

1 liter = 1,000 milliliters

M

mass 1.9 The amount of weight of an object in the metric system.

mean 10.2 The sum of a set of numbers divided by the number of addends.

median 10.2 The middle number when the data is listed in order.

meter (m) 1.9 The basic unit of length in the metric system.

metric system 1.9 A system of units of measure that uses prefixes. The basic unit of length is the meter, the basic unit of capacity is the liter, and the basic unit of mass is the gram.

metric ton (t) 1.9 A metric unit of weight.

1 metric ton = 1,000 kilograms

mile (mi) 6.11 A unit of length commonly used in the United States.

1 mile = 5,280 feet

milligram (mg) 1.9 A metric unit of weight.

1 milligram = 0.001 gram

milliliter (mL) 1.9 A unit of capacity in the metric system.

1 milliliter = 0.001 liter

millimeter (mm) 1.9 A unit of length in the metric system.

1 millimeter = 0.001 meter

mixed number 5.9 A number such as $1\frac{3}{4}$ that has a whole number part and a fraction part.

Glossary **379**

mode 10.2 The number or item that appears most often in a set of data. Some sets of data have no mode or more than one mode.

multiples of a number 6.3 Numbers that have the given number as a factor. The multiples are found by multiplying the number by 0, 1, 2, 3, and so on.

Multiplication Property of Equality 3.1 If you multiply by a quantity on one side of an equal sign, you must multiply by the same quantity on the other side to keep the two sides equal.

$$(3 \times 2) \times 5 = (2 \times 3) \times 5$$

N

negative number 12.1 A number that is less than zero.

net 9.9 A flat pattern that can be folded to make a solid.

numerator 5.5 In the fraction $\frac{3}{4}$, the numerator is 3. It tells the number of objects or parts being considered.

O

obtuse angle 8.2 An angle whose measure is between 90° and 180°.

obtuse triangle 8.3 A triangle that has one obtuse angle.

octagon 8.4 A polygon with eight sides.

ordered pair 10.9 A pair of numbers, such as (3, 2), used to locate a point on a map or grid. The first number tells the horizontal position, and the second number tells the vertical position.

order of operations 12.4 The order that operations should be performed in a mathematical expression. First, do operations in parentheses. Next, do all multiplication and division from left to right. Then, do all addition and subtraction from left to right.

ounce (oz) 6.11 A unit of weight in the customary system of measurement.

$$16 \text{ ounces} = 1 \text{ pound}$$

outcome 10.10 A possible result.

output 12.6 The number you get out of a function when you perform an operation.

P

parallel lines 8.1 Lines that lie in the same plane and do not intersect.

parallelogram 8.5 A quadrilateral with two pairs of parallel sides.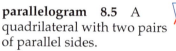

pentagon 8.4 A polygon with five sides.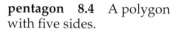

percent 11.6 A ratio that compares a number to 100. The symbol for percent is %.

perimeter 9.1 The distance around a polygon.

perpendicular lines 8.1 Lines that intersect at right angles.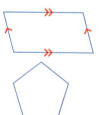

pi (π) 9.2 The quotient of the circumference of a circle divided by the diameter of the circle. An approximation for π is 3.14.

pint (pt) 6.11 A unit of capacity in the customary system of measurement.

$$1 \text{ pint} = 2 \text{ cups}$$

place-value position 1.1 The position of a digit in a numeral.

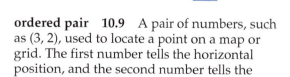

polygon 8.4 A closed figure in a plane made up of line segments that meet but do not cross.

positive numbers 12.1 Numbers that are greater than zero.

pound (lb) 6.11 A unit of weight in the customary system of measurement.

$$1 \text{ pound} = 16 \text{ ounces}$$

prime factorization 5.3 The way to express a composite number as the product of prime numbers.

$$60 = 2 \times 2 \times 3 \times 5$$

prime number 5.2 A whole number that has exactly two factors—itself and 1.

$$2 \quad 3 \quad 5 \quad 7 \quad 11$$

prism 8.11 A solid figure that has two congruent bases and parallelograms for faces.

probability 10.10 How likely or unlikely something is to happen.

product 3.1 The answer to a multiplication problem. In 6 × 6 = 36, the product is 36.

protractor 8.2 An instrument used to measure angles.

pyramid 8.11 A solid figure whose base can be any polygon and whose faces are triangles.

Q

quadrilateral 8.5 A polygon with four sides.

quart (qt) 6.11 A unit of capacity in the customary system of measurement.

$$1 \text{ quart} = 2 \text{ pints}$$

quotient 4.1 The answer to a division problem.

R

radius 8.7 A line segment from the center of a circle to any point on the circle.

range 10.1 The difference between the greatest and least number in a set of data.

rate 11.3 A comparison by a ratio of two quantities using different kinds of units.

ratio 11.1 A comparison of two numbers by division.

ray 8.1 An endpoint and all the points that continue forever in a certain direction from that endpoint. Rays are named by the endpoint and another point on the ray.

reciprocal 7.6 The product of a number and its reciprocal is 1.

$$\tfrac{3}{4} \times \tfrac{4}{3} = 1, \text{ so } \tfrac{3}{4} \text{ and } \tfrac{4}{3} \text{ are reciprocals.}$$

rectangle 8.5 A parallelogram with four right angles.

rectangular prism 8.11 A solid figure with six faces that are rectangles.

rectangular pyramid 8.11 A solid figure whose base is a rectangle and whose faces are triangles.

reflection 8.10 (flip) A figure is flipped over a line.

regular polygon 8.4 A polygon that has all sides and all angles congruent.

remainder 4.1 The amount left over after a division has been completed. The remainder must be less than the divisor.

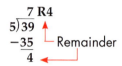

rhombus 8.5 A parallelogram with four sides the same length.

right angle 8.2 An angle that measures 90°.

right triangle 8.3 A triangle that has one right angle.

rotation 8.10 (turn) Turning a figure around a point.

S

sample 10.13 Part of a population that may represent the population at large.

scale 11.5 Drawing to the measurements of the real object.

scale drawing 11.5 A map or model of a real object that is enlarged or reduced by the same factor.

scalene triangle 8.3 A triangle with all sides of different lengths.

side of an angle 8.2 Either of the two rays forming an angle.

simplest form (of a fraction) 5.7 A fraction written so that the only common factor of the numerator and denominator is 1.

sphere 8.11 A solid figure that is shaped like a round ball.

square 8.4 A polygon with four right angles and four congruent sides.

standard form 1.2 The way a number is usually written. The standard form for two thousand forty-eight is 2,048.

statistics 10.1 The study of collecting, organizing, and interpreting data.

stem 10.4 The greatest place value common to all the data that is written to the left of the line in a stem-and-leaf plot.

stem-and-leaf plot 10.4 A frequency distribution that arranges data in order of place value. The last digit of the number is the *leaf* and the digits to the left of the leaf are the *stems*.

surface area 9.10 The total area of the surface of a solid.

symmetry 8.9 A figure has symmetry if a line can divide the figure into two equal parts.

T

tally mark 10.1 A mark that is made on a frequency table to show each time a piece of data occurs.

terms 11.1 The numbers compared in a ratio.

theoretical probability 10.10 The chance that an event will occur.

three-dimensional figure 8.11 A figure that does not lie in a plane.

ton (T) 6.11 A unit of weight in the customary system of measurement.

1 ton = 2,000 pounds

transformation 8.10 When you move a figure in geometry.

translation 8.10 (slide) Every point in the figure slides the same distance in the same direction.

trapezoid 8.5 A quadrilateral with exactly one pair of parallel sides.

tree diagram 10.12 A diagram that shows combinations of outcomes of an event.

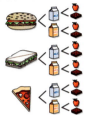

triangle 8.4 A polygon with three angles.

triangular prism 8.11 Prism whose bases are triangles.

triangular pyramid 8.11 Pyramid whose base is a triangle.

U

unit rate 11.3 Involves quantities that have different units of measurement and the denominator is always 1.

V

variable 12.2 A letter or symbol that represents a number in an algebraic expression.

Venn diagram 9.8 A diagram that uses a large rectangle with circles inside to represent a certain situation.

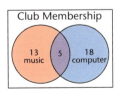

vertex (of an angle) 8.2 The common endpoint of the rays forming an angle.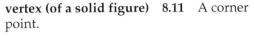

vertex (of a solid figure) 8.11 A corner point.

volume 9.11 The number of unit cubes needed to fill a solid figure.

W

weight 6.11 The attribute that tells how heavy an object is.

whole number 1.1 A positive number or zero that does not have a decimal part and is not part of a fraction.

Y

yard (yd) 6.11 A unit of length in the customary system of measurement.

1 yard = 3 feet

Z

Zero Property of Multiplication 3.1 If you multiply any number by zero, the product is zero.

8 × 0 = 0

0 × 8 = 0

Glossary 383

INDEX

A

Abacus, 3
Acute angles, 216–218
Acute triangle, 220–223
Addition
   angles, sum of measurement of, 220–223
   column addition, xxii–xxiii
   decimals, 44–45
   with fractions, 158–161, 166–167, 170–171
   magic squares, 315
   mental math, 34–37
   mixed numbers, 176–177
   order of operations, 356–357
   perimeter of polygons, 250–251
   properties, 30–33
   Property of Equality, 30–33
   statistics, 286–287
   whole numbers, xx–xxiii
Angles
   acute, obtuse, and right, 216–218
   congruent, 234–235
   estimating measures, 216–218
   measuring and drawing, 216–218
   naming, 216–218
   triangles, 220–221
Area
   of parallelograms, 260–262
   problem-solving application, 276–277
   of rectangles, 254–257
   surface area, 272–273
   of triangles, 265–266
Associative Property of Addition, 30–33
Associative Property of Multiplication, 60–62
Attic numerals, 15
Average, 286–287

B

Bar graphs, 292–293
Base (b)
   area of parallelograms, 260–262
   area of triangles, 264–266
   volume of rectangular prisms, 274–275
Base (exponents), 89
Brackets and order of operations, 356–357

C

Capacity, 20–22, 114–115
Center of circle, 230–232
Centimeter (cm), 20–22
   cubic centimeter (cm^3), 274–275
Chances (probability), 304–309
Charts and tables. *See also* Graphs.
   frequency charts and tables, 284–285
   function tables, 362–365
   ratio table, 322–323
Checkerboard
   problem-solving skills, 23
Chord, 230–232
Circle graphs, 338–339
Circles, 230–232
   center of, 230–232
   chord, 230–232
   circumference of, 252–253
   diameter of, 230–232
   pi (π) and circumference, 252–253
   radius of, 230–232
Circumference, 252–253
Column addition, xxii–xxiii
Common denominators
   adding fractions, 158–159
   comparing and ordering fractions, 164–165
   least common denominator, 162–163
   renaming sums, 178–179
   subtracting fractions, 158–159
   unlike denominators, 166–171
Commutative Property of Addition, 30–33
Commutative Property of Multiplication, 60–61
Comparing and ordering
   decimals, 16–17
   fractions, 164–165
   whole numbers, 6–8
Compatible numbers, 92–94
Composite numbers, 128–129
Cone, 240–242
Congruent figures, 234–235
Corresponding parts of congruent figures, 234–235
Cube, 240–242
Cubic centimeters (cm^3), 274–275
Cubic meters (m^3), 274–275

Cubic millimeters (mm³), 274–275
Cubic units of measurement, 274–275
Cups (c), 180–181
Customary units of measure, 180–181
   area of rectangles, 254–257
   capacity, 180–181
   computation, 182–183
   length, 180–181
   volume, 180–181
   weight, 180–181

D

Data and statistics, 284–285
   mean, median, and mode, 286–287
Decimal places, 14–15
Decimal points
   multiplication, 78–85
   multiplying and dividing by 10, 100, and 1,000, 78–79, 112–113
Decimals
   addition of, 44–45
   comparing and ordering, 16–17
   division of, 112–113, 116–117
   fractions as, 150–151
   multiplication of, 78–85
   place-value positions, 14–15
   quotients, 112–113, 116–117
   rounding, 18–19
   subtraction of, 46–47
   whole numbers, dividing decimals by, 116–117
   whole numbers, multiplying decimals by, 78–82
Decimeter (dm), 20–22
Degrees
   angle measurement, 216–218
Dekameter (dkm), 20–22
Denominators
   adding fractions with unlike, 166–167, 170–171
   like denominators, 158–159
   simplest form, fractions in, 138–139, 160–161
   subtracting fractions with unlike, 168–171
   unlike denominators, operations with, 166–171

Diagrams
   tree diagrams, 308–309
   Venn diagrams, 268–269
Diameter, 230–232
   and circumference, 252–253
Difference. See Subtraction.
Discounted prices, 342–343
Distances
   models, 328–329
   scale drawings, 328–329
Dividends. See Division.
Divisibility rules, 126–127
Division
   compatible numbers, 92–94
   decimals, 112–113, 116–117
   fact families, xxviii–xxix
   with fractions, 202–205
   fractions in simplest form, 160–161
   of greater numbers, 98–99, 110–111
   mean in statistics, 286–287
   by one-digit numbers, xxxii–xxxix
   order of operations, 356–357
   patterns, xxx–xxxi
   prime factorization, 130–131
   remainders, xxxii–xxxiii
   short division, 123
   statistics, 286–287
   ten, division by multiples of, 104–107
   by two-digit numbers, 108–111
   whole numbers, dividing decimals by, 116–117
   zeros in quotient, 96–97
Divisors. See Division.
Drawings
   problem-solving strategies, 228–229
   scale drawings, 328–329

E

Edge, 240–242
Elapsed time, 50–51
Eliminating possibilities, problem-solving strategies, 118–119
Equations, 354–355
   parentheses and order of operations, 356–357
Equilateral triangle, 220–223
Equivalent fractions, 134–137
   comparing and ordering fractions, 164–165
   in simplest form, 160–161

Equivalent ratios, 322–323
Estimation. *See* also Rounding.
    angles, estimating measures, 216–218
    choose exact answer or estimate, 52–53
    guess and check problem-solving strategy, 42
Expanded form of numbers, xvi–xvii, 4–5
Exponents, 89

F

Face, 240–242
Fact families, division and, xxviii–xxix
Factors
    common, 132–133
    greatest common, 132–133
    prime factors, 130–131
    related facts and missing factors, xxviii–xxix
    simplest form of fractions, 160–161
Factor tree, 130–131
Flipping shapes, 238–239
Foot (ft), 180–181
    cubic foot (ft^3), 274–275
    square foot (ft^2), 272–273
Four-step plan, problem-solving skill, 12–13
Fraction strips, 158, 160, 164, 168
Fractions
    addition of, 158–161, 166–167, 170–171
    comparing and ordering, 164–165
    as decimals, 150–151
    division of, 202–205
    equivalent fractions, 134–137
    greater than one, 142–144
    improper fractions, 142–144, 146–149
    least common denominator, 162–163
    multiplication of, 190–197
    number, fraction of a, 190–191
    prime factors, 130–131
    ratios, 318–320, 322–323
    renaming mixed numbers, 148–149
    simplest form, fractions in, 138–139, 160–161
    subtraction of, 158–161, 168–171
    whole numbers, multiplication with, 190–191
Frequency, 284–285
Frequency charts and tables, 284–285
Functions, 362–365

G

Gallons (gal), 180–181
Geometry
    area, 254–257, 260–262, 264–266
    circles, 230–232
    circumference, 252–253
    parallelograms, 226–227, 260–262
    perimeter, area, and volume, 276–277
    polygons, 226–227
    rectangles, 226–227
    squares, 226–227
    triangles, 220–223, 264–266
Grams (g), 20–22
Graphs
    bar graphs, 292–293
    circle graphs, 338–340
    line graphs, 294–295
    ordered pairs, 302–303
    problem-solving skills and strategies, 288–289, 298–299
Greater numbers
    division of, 98–99, 110–111
    subtraction of, 38–39
Greater than one, fractions, 142–144
Greater than symbol (>), 6–8
Greek Attic numerals, 15
Grids, 302–303
Group, fractions of parts of, xl–xli
Grouping
    Associative Property of Addition, 30–33
    Associative Property of Multiplication, 60–61
Guess and check problem-solving strategy, 42–43

H

Hectometer, 20–22
Height (*h*)
    area of parallelograms, 260–262
    area of triangles, 265–266
    surface area, 272–273
    and volume, 274–275
Hexagons, 224–225
History
    abacus, 3
    Attic numerals, 15
Horizontal bar graphs, 292–293

I

Identity Property of Addition, 30–33
Identity Property of Multiplication, 60–61
Improper fractions, 142–144, 146–149
    renaming, 146–147
Inches (in.), 180–181
    cubic inch (in.^{3}), 274–275
    square inch (in.^{2}), 254–255
Integers, 350–351
Isosceles triangle, 220–223

K

Kilograms (kg), 20–22
Kilometers (km), 20–22
    square kilometers (km^2), 254–255

L

Least common denominator, 162–163
Least common multiple, 162–163
Length (l), 20–22
    and area, 254–255
    computation, 182–183
    customary units, 180–181
    metric units, 20–22
    scale drawings, 328–329
    surface area, 272–273
Less than symbol (<), 6–8
Like denominators, 158–159
Line graphs, 294–295
Lines
    intersecting, 214–215
    optical illusions, 211
    parallel, 214–215
    perpendicular, 214–215
    segments, 214–215
    symmetry, line of, 236–237
Liters (L), 20–22
Location
    directions, models, 328–329
    graphing ordered pairs, 302–303
Logical reasoning, 198–199

M

Magic squares, 315
Maps, 328–329
Mass, measurement of, 20–22
Mean in statistics, 286–287
Measurement
    of angles, 216–218
    of capacity, 180–181
    of circumference, 252–253
    computation with measurements, 182–183
    of data, 286–287
    of perimeter of polygons, 250–251
    scale drawings, 328–329
    sum of measurement of angles, 220–221
    of surface area, 272–273
    units of measure, 114–115
    of weight, 180–181
Median in statistics, 286–287
Mental images, 165
Mental math
    addition, 34–37
    compatible numbers, 92–94
    dividing by multiples of ten, 104–105
    estimating products, 62–65
    estimating sums and differences, 36–37
    multiplication, 62–65
    multiplying and dividing by 10, 100, and 1,000, 62–63
    short division, 123
    subtraction, 34–37
Meters (m), 20–22
    cubic meters (m^3), 274–275
    square meters (m^2), 254–255
Metric units, 20–22
    area, 254–257, 260–263, 266–267
    capacity, 20–22
    computation, 180–181
    mass, 20–22
    renaming, 20–22, 114–115
Miles (mi), 180–181
Milliliters (mL), 20–22
Millimeters (mm), 20–22
    cubic millimeters (mm^3), 274–275
    square millimeters (mm^2), 254–255
Millions, place value, 4–5
Mixed numbers, 142–144, 146–149
    addition and subtraction, 176–177
    multiplication, 200–201
    renaming, 148–149

Mode in statistics, 286–287
Multiple solutions, problem solving, 48–49
Multiples of ten, division by, 104–107
Multiplication
    Associative Property of Multiplication, 60–61
    circumference of circles, 251–252
    Commutative Property of Multiplication, 60–61
    decimals, 78–85
    exponents, 89
    with fractions, 190–197
    hundreds, multiples of, 62–63, 70–71
    identity property, 60–61
    least common multiple, 162–163
    mental math, 62–65
    one, property of, 60–61
    by 1-digit numbers, xxvi–xxvii
    one hundred, multiples of, 62–63, 70–71
    one thousand, multiples of, 62–63
    order of operations, 356–357
    perimeter of polygons, 250–251
    properties of, 60–61
    tens, multiples of, 62–63, 70–71
    thousands, multiples of, 62–63
    by 3-digit numbers, 74–75
    by 2-digit numbers, 72–73
    of whole numbers, xxvi–xxvii, 62–75
    of whole numbers and fractions, 190–191
    whole numbers, multiplying decimals by, 78–82
    zero, property of, 60–61

N

Naming angles, 216–218
Naming polygons, 224–225
Nets, 270–271
90° angles, 216–218
Number lines, 350–351
Number puzzles, 77
Numerals
    Attic numerals, 15
Numerators
    comparing and ordering fractions, 164–165
    simplest form, fractions in, 160–161

O

Obtuse angles, 216–218
Obtuse triangles, 220–223
Octagons, 224–225
One, fractions greater than, 142–144
One, property of, 60–62
One-digit numbers
    division by, xxxii–xxxix
    multiplication by, xxvi–xxvii
Optical illusions, 211
Order and sequence
    comparing and, 6–8, 16–17
    operations, order of, 356–357
    parentheses, 356–357
Order property of addition (Commutative Property), 30–33
Order property of multiplication (Commutative Property), 60–61
Ordered pairs, 302–303
Ounces (oz), 180–181

P

Pairs
    ordered pairs, 302–303
Parallel lines, 214–215
Parallelograms, 226–227
    area of, 260–262
Parentheses and order of operations, 356–357
Patterns
    division patterns, xxx–xxxi
    mental images, 165
    pentominoes, 233
    problem-solving strategies, 76–77
Pentagons, 224–225
Pentominoes, 233
Percentages, 330–337
Perimeter, 250–251
Pi (π) and circumference, 252–253
Pint (pt), 180–181
Place value and place-value positions
    abacus, 3
    Attic numerals, 15
    decimals, 14–15
    renaming metric units, 20–22
    renaming whole numbers, 2–3
Points and optical illusions, 211

Polygons
    parallelograms, 224–225
    perimeter of, 250–251
    rectangles, 226–227
    regular, 224–225
    squares, 224–227
Possibilities, problem-solving strategies, 118–119
Pounds (lb), 180–181
Power (exponents), 89
Predictions
    probability, 304–309
    and statistics, 310–311
Prime factors and prime factorization, 130–131
Prime numbers, 128–129
Prisms, rectangular
    surface area, 272–273
    volume, 274–275
Probability, 304–307
Problem-solving applications
    choosing exact answer or estimate, 52–53
    discounts, 342–343
    four-step plan, 12–13
    graph, choosing the proper, 298–299
    remainders, 100–102
    using perimeter, area or volume, 276–277
Problem-solving strategies
    choosing method of computation, 140–141
    draw it, 258–259
    eliminating possibilities, 118–119
    equation writing, 366–367
    guess and check, 42–43
    logical reasoning, 198–199
    make a drawing, 228–229
    more than one solution, 48–49
    patterns, 76–77
    simpler problem, solving a, 206–207
    Venn diagrams, 268–269
    working backward, 358–359
Products. *See* Multiplication.
Properties of addition, 30–33
Properties of multiplication, 60–61
Protractor, 216–218
Puzzles
    length estimation, 119
    magic squares, 315
    number puzzles, 77
    tangrams, 155

Pyramids
    rectangular, 240–242
    triangular, 240–242

Q

Quadrilateral, 224–227
    parallelogram, 226–227
    rectangle, 226–227
    rhombus, 226–227
    square, 226–227
    trapezoid, 226–227
Quarts (qt), 180–181
Quotients. *See* Division.

R

Radius of circle, 230–232
    and circumference, 252–253
Range of data, 284–285
Ratio table, 322–323
Ratios, 318–320
    equivalent ratios, 322–323
    scale drawings, 328–329
Rays, 214–215
Rectangles
    area of, 254–257
    squares, 224–227
Rectangular prisms
    surface area, 272–273
    volume, 274–275
Reflection, 238–239
Related facts and missing factors, xxviii–xxix
Renaming numbers
    capacity, 180–181
    computation with measurements, 182–183
    fractions in simplest form, 138–139, 160–161
    length, 180–181
    metric units, 20–22
    mixed numbers, 148–149
    weight, 180–181
    whole numbers, 2–5
Right angles, 216–218
Right triangle, 220–223
Rhombus, 226–227
Roman numerals, 57
Rotation, 238–239

Rounding
  decimals, 18–19
  multiplication, 64–65
  whole numbers, xviii–xix, 10–11

S

Sale prices and discounts, 342–343
Samples and statistics, 310–311
Scale drawings, 328–329
Scientific notation, 347
Segments, 214–215
  congruent figures, 234–235
  optical illusions, 211
Shapes
  area of, 254–257, 260–262, 264–265
  circles, 230–232
  congruent figures, 234–235
  flipping, 238–239
  parallelograms, 226–227
  polygons, 224–225
  rectangles, 226–227
  sliding, 238–239
  squares, 226–227
  symmetry, 236–237
  tangrams, 155
  turning, 238–239
Short division, 123
Simpler problem, solving, 206–207
Simplest form, fractions in, 138–139, 160–161
Sliding shapes, 238–239
Solutions, more than one, 48–49
Square feet (ft^2), 254–255
Square inches (in.^{2}), 254–255
Square kilometers (km^2), 254–255
Square meters (m^2), 254–255
Square millimeters (mm^2), 254–255
Square units of measurement, 254–255
  surface area, 272–273
Square yards (yd^2), 254–255
Squares, 226–227
Standard form of numbers, xvi–xvii, 2–5
Statistics
  data and, 284–285
  frequency tables, 284–285
  mean, median, and mode, 286–287
  and predictions, 310–311
  problem-solving applications, 288–289, 298–299

Stem-and-leaf plot, 290–291
Subtraction
  data, range of, 284–285
  with decimals, 46–47
  with fractions, 158–161, 168–171
  of greater numbers, 38–39
  mental math, 34–37
  with mixed numbers, 176–177
  order of operations, 356–357
  renaming mixed numbers, 178–179
  whole numbers, xxiv–xxv, 38–39
Sum. *See* Addition.
Surface area, 272–273
Surveys and statistics, 284–285
Symmetry, 236–237

T

Tangrams, 155
10%, 334–335
Tesselation, 247
Three-dimensional figures, 240–242
  nets, 270–271
Time
  computation, 182–183
  elapsed time, 50–51
Tons
  customary units of measurement, 180–181
  metric units of measurement, 20–22
Transformations, 238–239
  reflection, 238–239
  rotation, 238–239
  translation, 238–239
Translation, 238–239
Trapezoid, 226–227
Tree diagrams, 308–309
  factor tree, 130–131
Triangles
  acute, 220–223
  angles of, 220–223
  area of, 264–266
  congruent figures, 234–235
  equilateral, 220–223
  isosceles, 220–223
  obtuse, 220–223
  right, 220–223
  scalene, 220–223
Triangular prism, 240–242

Triangular pyramid, 240–242
Turning shapes, 238–239
Two–digit numbers
    division by, 108–109
    multiplication by, 72–73

U

Units of measure
    area, 254–255, 260–262
    cubic units, 274–275
    square units, 272–273
    volume, 274–275
Unlike denominators
    adding fractions with, 166–167, 170–171
    subtracting fractions with, 168–169, 170–171

V

Venn diagrams, 268–269
Vertex, 240–242
Vertical bar graphs, 292–293
Volume (V), 274–275

W

Weight, 180–181
Whole, fractions of parts of, xl–xli
Whole numbers
    addition of, xx–xxiii
    and decimals, 116–117
    dividing decimals by, 116–117
    fractions, division with, 202–203
    fractions, multiplication with, 190–191, 196–197
    multiplication of, 64–75
    renaming, 2–5
    rounding, xviii–xix
    subtraction of, xxiv–xxv
Width (w)
    and area, 254–255
    surface area, 272–273
    and volume, 274–275
Working backward, problem-solving strategy, 358–359

Y

Yards (yd), 180–181
    cubic yards (yd^3), 274–275
    square yards (yd^2), 254–255

Z

Zero, Property of Addition, 30–33
Zero Property of Multiplication, 60–61